AF677968

Adaptation of Marine Animals to Extreme Environments

Guest Editor

Taewon Kim

MDPI

Basel • Beijing • Wuhan • Barcelona • Belgrade • Novi Sad • Cluj • Manchester

Guest Editor
Taewon Kim
Marine Zoology Lab.
Department of Ocean Sciences
Inha University
Incheon
Republic of Korea

Editorial Office
MDPI AG
Grosspeteranlage 5
4052 Basel, Switzerland

This is a reprint of the Special Issue, published open access by the journal *Journal of Marine Science and Engineering* (ISSN 2077-1312), freely accessible at: https://www.mdpi.com/journal/jmse/special_issues/cl_adaptation_marine_animals_extreme_environments.

For citation purposes, cite each article independently as indicated on the article page online and as indicated below:

Lastname, A.A.; Lastname, B.B. Article Title. *Journal Name* **Year**, *Volume Number*, Page Range.

ISBN 978-3-7258-5709-8 (Hbk)
ISBN 978-3-7258-5710-4 (PDF)
https://doi.org/10.3390/books978-3-7258-5710-4

© 2025 by the authors. Articles in this book are Open Access and distributed under the Creative Commons Attribution (CC BY) license. The book as a whole is distributed by MDPI under the terms and conditions of the Creative Commons Attribution-NonCommercial-NoDerivs (CC BY-NC-ND) license (https://creativecommons.org/licenses/by-nc-nd/4.0/).

Contents

About the Editor

Taewon Kim

Taewon Kim is currently an Associate Professor in the Department of Ocean Sciences at Inha University. He also serves as an Adjunct Professor in the Program in Biomedical Science and Engineering at the same university. He earned his Ph. D. in behavioral ecology of fiddler crabs from Seoul National University. During his postdoctoral period at the Stanford University and Monterey Bay Aquarium Research Institute, he worked on various projects interweaving the following keywords: climate change, marine bioinvasion, ocean acidification, and deep-sea animals. As a senior research scientist at the Korea Institute of Ocean Science and Technology, Dr. Kim collaborated with scientists and engineers to address environmental challenges including those in extreme environments such as the Antarctic Ocean and deep-sea hydrothermal vents. Since 2017, he has been leading the Marine Zoology Laboratory at Inha University where he enjoys conducting diverse projects on marine animals with his students and collaborators. His primary research focus is on anthropogenic influences (e.g., climate change, plastic pollution, and vibrational/noise pollution) on the marine ecosystems as well as the adaptation of marine animals to extreme environments. He believes that how marine animals adapt to extreme environments can provide insights for the development of new technology.

Journal of
Marine Science and Engineering

MDPI

Editorial

Adaptation of Marine Animals to Extreme Environments

Taewon Kim [1,2]

[1] Marine Zoology Laboratory, Department of Ocean Sciences, Inha University, Incheon 22212, Republic of Korea; ktwon@inha.ac.kr
[2] Program in Biomedical Science & Engineering, Inha University, Incheon 22212, Republic of Korea

1. Introduction

The Earth's oceans are vast, mysterious, and replete with environments that test the limits of biological survival. Among these, extreme marine environments such as the polar regions, deep-sea abyssal plains, and hydrothermal vent systems stand out for their challenging physical and chemical conditions. Temperatures plunge below freezing in polar waters, hydrostatic pressures in the deep sea crush unprotected structures, and hydrothermal vents spew fluids exceeding 350 °C loaded with toxic chemicals. Despite these formidable obstacles, marine organisms have evolved extraordinary physiological, biochemical, and behavioral adaptations, not only to survive but also to thrive and reproduce [1,2].

Understanding how marine animals adapt to such extremes is not merely a matter of academic interest. These adaptations provide insights into evolutionary resilience, the limits of life, and the potential for life in extraterrestrial settings [2]. Moreover, in an era of accelerating climate change, understanding the mechanisms that underlie resilience and adaptation becomes crucial for forecasting the responses of marine ecosystems to global perturbations [3,4].

This Special Issue brings together seven studies that collectively deepen our understanding of adaptation strategies employed by marine organisms inhabiting extreme environments. The contributions span polar ecosystems, deep-sea habitats, hydrothermal vent communities, and rare pelagic intrusions, employing a combination of molecular, physiological, and ecological approaches. In the following sections, we synthesize the main findings of each paper and contextualize them within the broader landscape of marine adaptation research.

2. Adaptations in Polar Environments

2.1. Antarctic Amphipods: Stability Amid Change

Ahn et al. (2022) provide a rare glimpse into the trophic ecology of *Gondogeneia antarctica*, a gammarid amphipod inhabiting the rapidly warming fjord systems of the West Antarctic Peninsula. Unlike krill, which dominate much of Antarctic research, amphipods have received comparatively little attention despite their ecological importance [5]. This year-long study in Marian Cove employed stable isotope ($\delta^{13}C$ and $\delta^{15}N$) and fatty acid profiling to examine seasonal dietary shifts.

Surprisingly, the authors found minimal seasonal variation in the amphipod's diet. Fatty acid composition remained relatively stable between spring-summer (64%) and fall-winter (58%) periods, and isotopic shifts in $\delta^{13}C$ were modest. The data suggest a consistent reliance on benthic primary producers, particularly red algae and diatoms, throughout the year. This dietary stability in the face of seasonal change highlights the species' adaptation to a relatively predictable benthic environment and underscores its potential role in energy transfer within Antarctic coastal ecosystems.

The implications of this study are far-reaching. As global temperatures rise and seasonal cycles become more erratic, organisms with stable dietary patterns may be better positioned to endure environmental shifts [3,5]. Their foraging behavior was also found to change in response to other factors, such as decreased pH and salinity [5]. However, dependence on benthic sources also renders them vulnerable to bottom-up changes in the ecosystem, such as declines in benthic primary production due to increased glacial input or sedimentation.

2.2. Limpet-Algae Interactions: Parasitism or Mutualism?

Cho et al. (2022) explored the interaction between the Antarctic limpet *Nacella concinna* and its epibiotic partner, the coralline algae *Clathromorphum obtectulum*. Through a combination of laboratory experiments and field observations, the study painted a complex picture of symbiosis under environmental stress.

Limpets with algal cover exhibited higher mortality and heavier shells but showed no significant change in condition factor. Fatty acid analyses revealed an increase in saturated and a decrease in polyunsaturated fatty acids, indicating altered lipid metabolism. Interestingly, limpets with algae experienced less shell surface erosion, likely due to protection from endolithic boring organisms. The study concluded that while the interaction is largely parasitic under current conditions, it could shift toward mutualism as ocean temperatures and acidification progress.

This nuanced perspective is critical for understanding how species interactions may be reshaped by climate change [3]. Previous research has shown that *N. concinna* shells are susceptible to corrosion in seawater with reduced pH or salinity—conditions that may be caused by glacial meltwater input [6]. However, compensatory mechanisms may exist. For example, in the Antarctic clam *Laternula elliptica*, shell dissolution under reduced pH was mitigated by thickening of the periostracum layer [7]. These findings highlight the importance of investigating the plasticity and dynamic nature of symbioses to better predict species adaptation in a rapidly changing climate.

3. Hydrothermal Vent Ecosystems: Dispersal, Genomics, and Evolution

3.1. Vertical Dispersal of Vent Larvae

Minju Kim et al. (2022) tackled one of the enduring questions in deep-sea biology: how are isolated hydrothermal vent communities connected? Using MOCNESS net tows above vent fields in the Central Indian Ridge, they identified a diverse assemblage of gastropod and bivalve larvae in the upper 200 m of the water column. Many of these larvae exhibit morphological traits consistent with those of hydrothermal vent fauna, including species resembling *Bathymodiolus* spp.

The findings suggest that vent-associated larvae are capable of vertical migration from depths of ~2000 m, challenging the notion that vent systems are isolated biological islands. Such vertical transport likely enhances genetic connectivity and colonization potential across disjointed vent habitats, offering a mechanistic explanation for observed biogeographic patterns [1,8].

This discovery is vital for conservation biology. If vent larvae can indeed travel vertically and disperse over large distances, then marine protected areas for vent ecosystems must be designed with connectivity in mind [8].

3.2. Range Expansion and Genetic Continuity

In a complementary study, Hwang et al. (2022) reported the first discovery of the squat lobster *Munidopsis lauensis* in the Indian Ocean. Previously thought to be restricted to the southwest Pacific, the Indian Ocean specimens showed a 99.4–100% genetic match with Pacific populations based on mitochondrial COI barcoding. Phylogenetic divergence dating estimated the split to have occurred around 15 million years ago.

Structural analysis of COX1 proteins revealed conserved heme-binding regions, suggesting functional stability of oxygen-binding mechanisms [1]. These results imply a previously unrecognized biogeographic dispersal corridor between the Pacific and Indian Oceans, possibly facilitated by larval drift along mid-ocean ridges.

The broader implications extend to paleobiogeography and plate tectonics. If vent species can disperse across vast oceanic expanses, then the history of deep-sea fauna may be more continuous and interconnected than previously assumed [8].

3.3. Genomic Adaptations in Parasitic Polychaetes

Choi et al. (2022) focused on the deep-sea parasitic polychaete *Branchipolynoe onnuriensis*, which inhabits hydrothermal vent bivalves. Using whole-genome sequencing, the authors identified seven chitin synthase (CHS) genes, which were classified into two Type 1 and four Type 2 subgroups (A–D). Phylogenetic analysis revealed lineage-specific gene expansions, particularly in Group C.

Conserved motifs such as "EDR" and "QRRRW" were found across CHS genes, indicating functional conservation. The expansion and diversification of CHS genes may be key to the worm's adaptation to its parasitic lifestyle in chemically harsh environments. This study represents the first genomic insight into CHS diversity among parasitic annelids in hydrothermal ecosystems, highlighting molecular pathways of adaptation. Further research on CHS genes can also reveal the exceptional properties of exoskeletons of crustaceans found in the deep-sea hydrothermal vents [9–12].

4. Deep-Sea Adaptation and Pelagic Intrusion

4.1. Deep-Sea Pressure Adaptation: The Role of TMAO

Liu et al. (2022) explored how two hadal amphipods—*Hirondellea gigas* and *Alicella gigantea*—adapt to crushing deep-sea pressures. They measured concentrations of trimethylamine (TMA) and its oxidized form, trimethylamine N-oxide (TMAO), across eight tissues and compared them with those of the shallow-water shrimp *Penaeus vannamei*.

TMAO levels were significantly elevated in the hadal species, especially in external tissues such as the eyes and exoskeleton. A strong positive correlation was observed between TMA and TMAO in hadal species, but not in the shallow-water control, suggesting a pressure-regulated transformation process. Furthermore, the FMO3 enzyme responsible for this conversion displayed positively selected mutations in *A. gigantea*, potentially enhancing its catalytic efficiency under extreme pressure. These findings confirm TMAO's critical role in maintaining protein structure and cellular function in deep-sea environments [2,13].

This research also holds relevance for hydrothermal vent ecosystems, which are likewise situated in the deep sea. To understand the adaptations of marine animals inhabiting hydrothermal vents, comparative studies that decouple the influences of pressure and temperature are essential. Investigating key physiological and molecular features such as TMAO accumulation and chitin synthase (CHS) gene expression across both hadal and vent-endemic species could provide new insights into how marine organisms adapt to the specific challenges of deep-sea environments [10,12].

4.2. Pelagic Intrusion of Deep-Sea Species: A Case of Biogeographic Shift

Sung Kim et al. (2022) reported the first detection of the oceanic lightfish *Vinciguerria nimbaria* in Korean waters, based on high-throughput sequencing of 266 mixed fish egg samples collected from 78 coastal stations. COI barcoding and phylogenetic analysis identified 20 eggs of *V. nimbaria* across six samples.

Previously unrecorded in the region, *V. nimbaria* is a tropical mesopelagic species known for its diel vertical migrations and the formation of dense nighttime schools. Its presence in Korean waters may signal a northward range expansion potentially driven by ocean warming. This study exemplifies the power of DNA metabarcoding in detecting rare or cryptic species, highlighting the importance of molecular surveillance in tracking biogeographic shifts in response to climate change.

5. Conclusions

Together, these seven contributions underscore the remarkable adaptability of marine organisms to some of the most extreme environments on Earth. From the dietary stability of Antarctic amphipods and the complex symbiosis between limpets and algae, to the vertical dispersal of hydrothermal vent larvae and the molecular intricacies of chitin synthesis and pressure adaptation, each study adds a valuable piece to the puzzle of life under extreme conditions.

Several overarching themes emerge. First, molecular tools such as DNA barcoding, high-throughput sequencing, and genome analysis are revolutionizing our ability to detect, identify, and understand marine life in hard-to-access environments. Second, larval dispersal appears to be a key mechanism connecting isolated populations, facilitating gene flow and biogeographic expansion. Third, the impacts of climate change are not uniform; they reshape species interactions, distributions, and potentially even evolutionary trajectories [3].

Furthermore, the ecological and biogeographic revelations presented in this Special Issue have practical implications for marine conservation. As climate change accelerates, understanding the mechanisms of adaptation will be essential for designing effective marine protected areas, managing fisheries, and preserving biodiversity hotspots. In sum, the adaptations described in this Special Issue not only illuminate the resilience of marine life but also serve as sentinels of ecological change.

Funding: This research received no external funding.

Conflicts of Interest: The author declares no conflicts of interest.

List of Contributions:

1. Ahn, I.-Y.; Elias-Piera, F.; Ha, S.-Y.; Rossi, S.; Kim, D.-U. Seasonal Dietary Shifts of the Gammarid Amphipod *Gondogeneia antarctica* in a Rapidly Warming Fjord of the West Antarctic Peninsula. *J. Mar. Sci. Eng.* **2021**, *9*, 1447. https://doi.org/10.3390/jmse9121447.
2. Kim, M.; Kang, J.-H.; Kim, D. Holoplanktonic and Meroplanktonic Larvae in the Surface Waters of the Onnuri Vent Field in the Central Indian Ridge. *J. Mar. Sci. Eng.* **2022**, *10*, 158. https://doi.org/10.3390/jmse10020158.
3. Hwang, H.-s.; Cho, B.; Cho, J.; Park, B.; Kim, T. New Record of Hydrothermal Vent Squat Lobster (*Munidopsis lauensis*) Provides Evidence of a Dispersal Corridor between the Pacific and Indian Oceans. *J. Mar. Sci. Eng.* **2022**, *10*, 400. https://doi.org/10.3390/jmse10030400.
4. Liu, Q.; Jiang, S.; Li, W.; Pan, B.; Xu, Q. Trimethylamine N-Oxide (TMAO) and Trimethylamine (TMA) Determinations of Two Hadal Amphipods. *J. Mar. Sci. Eng.* **2022**, *10*, 454. https://doi.org/10.3390/jmse10040454.

5. Cho, B.; Bae, H.; Kim, T. The Symbiotic Relationship between the Antarctic Limpet, *Nacella concinna*, and Epibiont Coralline Algae. *J. Mar. Sci. Eng.* **2022**, *10*, 496. https://doi.org/10.3390/jmse10040496.
6. Choi, H.; Kim, S.L.; Jeong, M.-K.; Yu, O.H.; Eyun, S. Identification and Phylogenetic Analysis of Chitin Synthase Genes from the Deep-Sea Polychaete *Branchipolynoe onnuriensis* Genome. *J. Mar. Sci. Eng.* **2022**, *10*, 598. https://doi.org/10.3390/jmse10050598.
7. Kim, S.; Chin, B.-S.; Wang, S.-Y. Evidence of the Intrusion of the Oceanic Lightfish (*Vinciguerria nimbaria*) into Korean Waters Based on High-Throughput Sequencing of Mixed Fish Eggs. *J. Mar. Sci. Eng.* **2023**, *11*, 257. https://doi.org/10.3390/jmse11020257.

References

1. Van Dover, C.L. *The Ecology of Deep-Sea Hydrothermal Vents*; Princeton University Press: Princeton, NJ, USA, 2000.
2. Somero, G.N. Biochemical ecology of deep-sea animals. *Experientia* **1992**, *48*, 537–543. [CrossRef] [PubMed]
3. Levin, L.A.; Le Bris, N. The deep ocean under climate change. *Science* **2015**, *350*, 766–768. [CrossRef] [PubMed]
4. Thurber, A.R.; Sweetman, A.K.; Narayanaswamy, B.E.; Jones, D.O.; Ingels, J.; Hansman, R.L. Ecosystem function and services provided by the deep sea. *Biogeosciences* **2014**, *11*, 3941–3963. [CrossRef]
5. Park, S.; Ahn, I.-Y.; Sin, E.; Shim, J.H.; Kim, T. Ocean freshening and acidification differentially influences mortality and behavior of the Antarctic amphipod *Gondogeneia antarctica*. *Mar. Environ. Res.* **2020**, *154*, 104847. [CrossRef] [PubMed]
6. Sin, E.; Ahn, I.; Park, S.; Kim, T. Effects of low pH and low salinity induced by meltwater inflow on the behavior and physical condition of the Antarctic limpet, *Nacella concinna*. *J. Mar. Sci. Eng.* **2020**, *8*, 822. [CrossRef]
7. Seo, H.; Cho, B.; Joo, S.; Ahn, I.; Kim, T. Archival records of the Antarctic clam shells from Marian Cove, King George Island suggest a protective mechanism against ocean acidification. *Mar. Pollut. Bull.* **2024**, *200*, 116052. [CrossRef] [PubMed]
8. Vrijenhoek, R.C. Genetic diversity and connectivity of deep-sea hydrothermal vent metapopulations. *Mol. Ecol.* **2010**, *19*, 4391–4411. [CrossRef] [PubMed]
9. Cho, B.; Kim, D.; Bae, H.; Kim, T. Unique characteristics of the exoskeleton of *Austinograea rodriguez* in the Indian Ocean hydrothermal vent (Onnuri Vent Field). *Integr. Comp. Biol.* **2020**, *60*, 24–32. [CrossRef] [PubMed]
10. Cho, B.; Kim, D.; Kim, T. Exceptional properties of hyper-resistant armor of a hydrothermal vent crab. *Sci. Rep.* **2022**, *12*, 11816. Available online: https://www.nature.com/articles/s41598-022-15982-1 (accessed on 27 August 2025). [CrossRef] [PubMed]
11. Cho, B.; Seo, H.; Hong, J.; Jang, S.; Kim, T. Exoskeletal trade-off between claws and carapace in deep-sea hydrothermal vent decapod crustaceans. *Integr. Comp. Biol.* **2024**, *64*, 80–91. [CrossRef] [PubMed]
12. Cho, B.; Jang, S.-J.; Hwang, H.-S.; Kim, T. Convergent evolution of armor: Thermal resistance in deep-sea hydrothermal vent crustaceans. *Biology* **2024**, *13*, 956. [CrossRef] [PubMed]
13. Yancey, P.H.; Gerringer, M.E.; Drazen, J.C.; Rowden, A.A.; Jamieson, A. Marine fish may be biochemically constrained from inhabiting the deepest ocean depths. *Proc. Natl. Acad. Sci. USA* **2014**, *111*, 4461–4465. [CrossRef] [PubMed]

Disclaimer/Publisher's Note: The statements, opinions and data contained in all publications are solely those of the individual author(s) and contributor(s) and not of MDPI and/or the editor(s). MDPI and/or the editor(s) disclaim responsibility for any injury to people or property resulting from any ideas, methods, instructions or products referred to in the content.

Journal of
Marine Science and Engineering

MDPI

Article

Seasonal Dietary Shifts of the Gammarid Amphipod *Gondogeneia antarctica* in a Rapidly Warming Fjord of the West Antarctic Peninsula

In-Young Ahn [1,*], Francyne Elias-Piera [1,†], Sun-Yong Ha [1], Sergio Rossi [2,3] and Dong-U Kim [1,4]

1. Division of Ocean Sciences, Korea Polar Research Institute, 26 Songdomirae-ro, Incheon 21990, Korea; Francyne.ep@gmail.com (F.E.-P.); syha@kopri.re.kr (S.-Y.H.); kimduocean@kopri.re.kr (D.-U.K.)
2. DiSTeBA, University of Salento, Via Monteroni s/n, 73100 Lecce, Italy; sergio.rossi@unisalento.it
3. Instituto de Ciências do Mar-LABOMAR, Universidade Federal do Ceará, Av. da Abolição, Fortaleza 3207, Brazil
4. School of Earth and Environmental Sciences, Seoul National University, Seoul 08826, Korea

* Correspondence: iahn@kopri.re.kr

† Present address: Instituto Oceanográfico, Universidade de São Paulo, São Paulo 05508-900, Brazil.

Abstract: The amphipod *Gondogeneia antarctica* is among the most abundant benthic organisms, and a key food web species along the rapidly warming West Antarctic Peninsula (WAP). However, little is known about its trophic strategy for dealing with the extreme seasonality of Antarctic marine primary production. This study, using trophic markers, for the first time investigated seasonal dietary shifts of *G. antarctica* in a WAP fjord. We analyzed $\delta^{13}C$ and $\delta^{15}N$ in *G. antarctica* and its potential food sources. The isotopic signatures revealed a substantial contribution of red algae to the amphipod diet and also indicated a significant contribution of benthic diatoms. The isotope results were further supported by fatty acid (FA) analysis, which showed high similarities in FA composition (64% spring–summer, 58% fall–winter) between *G. antarctica* and the red algal species. *G. antarctica* $\delta^{13}C$ showed a small shift seasonally (−18.9 to −21.4‰), suggesting that the main diets do not change much year-round. However, the relatively high $\delta^{15}N$ values as for primary consumers indicated additional dietary sources such as animal parts. Interestingly, *G. antarctica* and its potential food sources were significantly enriched with $\delta^{15}N$ during the fall–winter season, presumably through a degradation process, suggesting that *G. antarctica* consumes a substantial portion of its diets in the form of detritus. Overall, the results revealed that *G. antarctica* relies primarily on food sources derived from benthic primary producers throughout much of the year. Thus, *G. antarctica* is unlikely very affected by seasonal Antarctic primary production, and this strategy seems to have allowed them to adapt to shallow Antarctic nearshore waters.

Keywords: *Gondogeneia antarctica*; seasonal dietary shift; macroalgae; benthic diatoms; C and N stable isotopes; West Antarctic Peninsula; King George Island; Marian Cove (62°13′ S; 58°47′ W)

1. Introduction

Antarctic marine ecosystems are rapidly changing, particularly along the West Antarctic Peninsula (WAP), due to regional warming and glacier melt [1–4]. The changes are most prominent in shallow coastal waters where marine-terminating glaciers have been rapidly retreating over the last half-century [5]. Aside from the ongoing climate impacts, ice disruptions are prevalent year-round in the intertidal and shallow subtidal zones, and benthic communities end up with low abundance and low diversity [6–8].

Amphipods are one of the few benthic fauna frequently observed in highly disturbed shallow habitats. Vagile amphipods aggregate in gravel or macroalgal beds, utilizing them as shelter and/or potential food sources [9–15]. Amphipod crustaceans are among the most specious fauna, with more than 800 species in the Southern Ocean; >500 are strictly endemic to the Antarctic [16] and remarkably abundant along the Antarctic

Peninsula [9,17,18]. Amphipods also show a highly diverse lifestyle associated with a variety of food sources and habitats [13,19–21].

The gammaridean amphipod *Gondogeneia antarctica* (Chevreux, 1906) is one of the most abundant species, particularly in shallow (<40 m) nearshore habitats, and is usually associated with macroalgal beds, which frequently form dense canopies in shallow waters along the WAP [9,11,22–26]. Nearshore Antarctic benthic communities rely on various food sources obtained through multiple pathways from both pelagic and benthic primary producers [27–31]. *G. antarctica* feeds primarily on benthic diatoms and macroalgae [10,13,32,33], although its gut contents indicated that it also consumes some animal parts (e.g., crustacean parts and sponge spicules) [9,10,13,34]. In turn, this species is preyed upon by various benthic invertebrates and demersal fishes, among other taxonomic groups [26,35–40]. Thus, *G. antarctica* plays an important role in shallow Antarctic marine food webs [40]. Therefore, the trophic response of *G. antarctica* to climate change and consequent constraints may be a good indication of the adaptability of this species, as well as of trophically related key organisms.

However, little is known about the trophic strategy of *G. antarctica* for dealing with the extreme seasonality in primary production of the Antarctic marine environment, and almost all feeding studies have been carried out during the summer when food availability is highest. A prerequisite to evaluate the adaptations or constraints of a key species to climate change is to understand its seasonal trophic characteristics (e.g., feeding preferences and the ability to mobilize energy during food-constrained periods). The Antarctic marine environment is characterized by small seasonal variations in seawater temperature and salinity, whereas primary phytoplankton production is extremely seasonal with explosive blooms occurring during a relatively short time in summer [41].

In this study, using trophic markers, we for the first time investigated dietary shifts in *G. antarctica* over 1 year in Marian Cove, a typical nearshore habitat for this species in the WAP. We analyzed carbon and nitrogen stable isotopes (SIs) and fatty acids (FAs) of *G. antarctica* along with its potential food sources (macroalgae, benthic diatoms, and suspended particulate organic matter (SPOM)), and examined any seasonal shifts in their trophic signatures. The findings from this study will enhance our understanding of the *G. antarctica* trophic strategy to deal with extremely seasonal food availability in its habitat and provide insight into likely future changes in the trophic relationships of nearshore benthic communities in the WAP.

2. Materials and Methods

2.1. Study Area

The King Sejong Station (62°13′ S, 58°47′ W) is located in Marian Cove, which is a glacial embayment (4.5 km long, ~1.5 km wide, ~130 m deep) within Maxell Bay bounded by the Weaver Peninsula and Barton Peninsula (Figure 1). Surface seawater freezes frequently in winter, although not every year, with variable cover of drifting ice throughout much of the year. Seawater temperature varies seasonally, from a maximum of ca. +1.5 °C in February to a minimum of ca. −1.8 °C in August; salinity varies much less, from 33.8 to 34.1 psu [42]. Further details of the hydrographic features and other environmental conditions of this area have been described elsewhere [43,44].

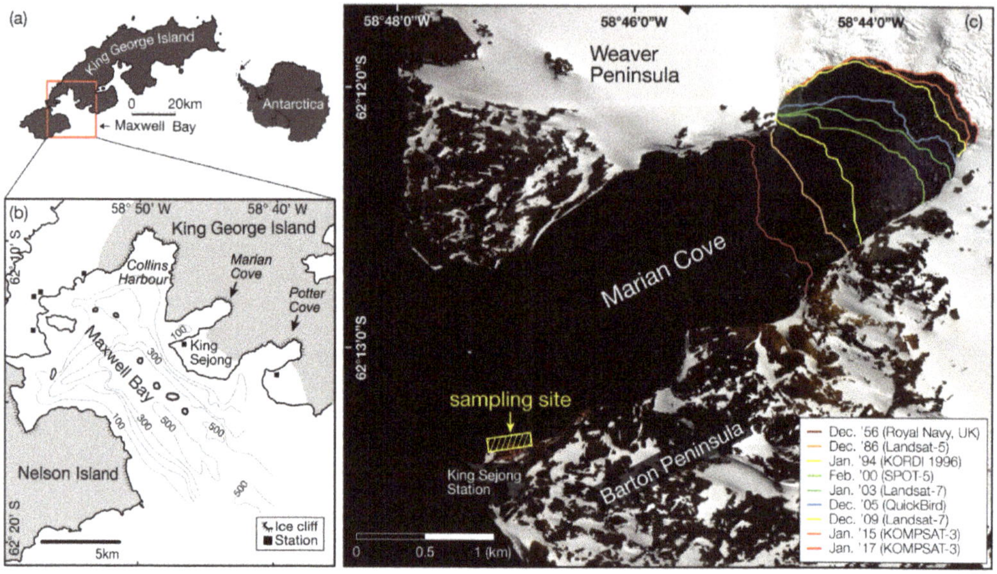

Figure 1. Study area. (**a**) Location of King George Island and Maxwell Bay. (**b**) Bathymetry of Maxwell Bay and its tributary embayments. Bathymetric contours are drawn based on information from the Atlas Hidrografico Chileno Antarctica of the Instituto Hidrografico de la Armada, Chile (1982). The gray area indicates glacial cover. (**c**) The sampling site in Marian Cove (hatched area) and glacier lines since 1956. The figure was adapted from [45] and the glacier lines were updated to December 2017.

Tidewater glaciers are well developed along the fringe of the inner cove, and the glaciers retreated by approximately 1.7 km from 1956 to 2013 [45]. Recent satellite images show that these glaciers have been continuously retreating since then [42]. Glacial retreat is accompanied by a series of processes (e.g., glacier calving and melting), introducing large volumes of ice, fresh meltwater, and terrigenous sediment to the cove, and eventually generating distinct environmental gradients along with glacial runoff [44,46–49]. Recent studies conducted in the cove have shown that these processes following the glacial retreats have impacted the marine inhabitants, particularly the benthic communities [42,45,49]. These previous studies have suggested Marian Cove as a model ecosystem for climate-related studies.

2.2. Selection of Potential Food Items for Analysis

G. antarctica has been known to predominantly rely on macroalgae and microphytobenthos for its main diets [10,13,32,33], despite the presence of a diverse array of food items (e.g., crustacean parts and sponge spicules) in its gut content [9,10,13,34]. Macroalgae, benthic diatom, and suspended particulate organic matter (SPOM) were selected primary food sources of *G. antarctica* for this study.

Four species of red algae (*Iridaea cordata, Palmaria decipiens, Gigartina skottsbergii,* and *Curdiea racovitzae*) and two species of brown algae (*Phaeurus antarcticus* and *Desmarestia* sp.), which are very common in this area [50–53], were collected. The red algae are known to be palatable to many herbivores, detritivores, and omnivores [10,34,54–56]. Benthic diatoms were reported as a primary food source for *G. antarctica* and other shallow-water amphipod species during austral summer [13,57]. Previous studies in this area have reported widespread and intense benthic diatom blooms in shallow waters, and the benthic diatoms were consumed by a variety of common benthic fauna [30,58]. Other studies in the peninsula region also reported that benthic diatoms associated with macroalgae were consumed selectively or randomly by herbivores, such as amphipods [13,59].

In addition to macroalgae and benthic diatom, SPOM was also selected as a potential food source, as it likely reflects organic detritus on the bottom, which are available to the amphipods in the study area. In general, SPOM, the most important form of organic particulate matter in the water column, is indicative of pelagic primary production. SPOM eventually settles to the bottom after bloom and constitutes part of sedimentary organic deposit, which is available to benthic fauna. In shallow nearshore waters, however, organic detritus settled on the bottom can be easily resuspended by tides or waves. Previous studies in this area reported that a substantial portion of SPOM in the shallow nearshore waters (<30 m) was frequently derived from benthic sources, such as benthic diatoms detached from their biotic and abiotic substrates and resuspended by wind-generated vertical mixing [60]. SPOM $\delta^{13}C$ values reported in a previous study in the same area during austral summer (-24.1 ± 0.2‰, $n = 6$) [61] were even close to the benthic diatom values (-23.3 ± 0.3‰ in [30]), while the reported SPOM δ13C values derived from phytoplankton ranged from -28.0 to -30.4‰ [59]. Thus, findings from previous studies suggest that SPOM in shallow nearshore Antarctic waters could consist of a mixture of organic matter derived from various sources resuspended from the sea floor, and it reflect organic detritus on the bottom available to the amphipods.

2.3. Sample Collection and Processing

G. antarctica, SPOM, and macroalgae were collected between January and November 2015 from intertidal and shallow subtidal waters (<0.5 m below low tide level) adjacent to the station. To obtain the SPOM, surface seawater was collected using 2 L sterile LDPE collapsible water bottles (DAIHAN, Wonju, South Korea), and the collected water was filtered (300–500 mL for each $\delta^{13}C$ replicate and 700–1000 mL for each $\delta^{15}N$ replicate) using a vacuum pump onto pre-combusted (450 °C, 4 h) glass fiber filters (25-mm in diameter; Whatman, Florham Park, NJ, USA). The filters were immediately frozen.

Macroalgae stranded on the shore were randomly collected by hand in February, March, May, and October but not in winter (June–September), during which the shallow seabed was covered by pack ice. The collected macroalgae seemed to have been recently detached by waves and were fresh at the time of collection onshore. The collected macroalgal fronds were washed with filtered seawater. All visible encrusted fauna and flora were removed from the surface and placed into plastic bags after being cut into pieces, and then frozen. *G. antarctica* was randomly collected using a hand net (20 cm diameter, 100 μm mesh) in the shallow subtidal zone (<50 cm). The collected SPOM, macroalgae, and animals were stored at −80 °C for up to 9 months at the station and then transported to the KOPRI for analysis. Isotopic data for benthic diatoms and additional data for macroalgae were obtained from a previous study performed in this area [30].

2.4. Stable Isotope Analysis (SIA)

The macroalgae and amphipods were freeze-dried and ground with a mortar and pestle into a homogenous powder at the KOPRI. Aliquots (0.60–0.80 mg each) of the homogenized samples were used for the SIA. The amphipod samples for each sampling time included both juveniles (body length < 0.9 cm) and adults (>0.9 cm) [10]. Among them the samples with body size (0.5–1.5 cm) were pooled to ensure a sufficient amount for reliable and representative composite samples.

The samples for the organic carbon isotope analysis were rapidly acidified with two or three drops of 1 N HCl to remove inorganic carbonates, and rinsed with Milli-Q water [62]. Because acid washing can affect the nitrogen isotope ratios of organic material [63], samples for the nitrogen isotope analysis were not acidified. Lipids were removed using the chloroform/methanol procedure outlined in [64]. The $\delta^{13}C$ and $\delta^{15}N$ analyses were performed with an isotopic mass spectrometer (Isoprime 100; Elementar, Manchester, UK) coupled with an elemental analyzer (Euro EA3028; EuroVector, Milan, Italy) following the procedure in [30].

The isotopic ratios were expressed in parts per thousand (‰) according to the following equation:

$$\delta X = [(R_{sample}/R_{standard}) - 1] \times 10^3$$

where X is ^{13}C or ^{15}N and R is the corresponding $^{13}C/^{12}C$ or $^{15}N/^{14}N$ ratio. The $R_{standard}$ values for ^{13}C and ^{15}N are from V-PDB and atmospheric N_2, respectively.

2.5. Fatty Acid (FA) Analysis in *Gondogeneia antarctica* and Red Algae

FAs were determined using a protocol slightly modified from those used in previous studies [65–67]. Aliquots (approximately 15 mg) of freeze-dried and homogenized amphipod and red algal samples were extracted with a mixture of dichloromethane: methanol (3:1, *v:v*). The extract was separated into three fractions (neutral lipids, free fatty acids (FFAs), and polar lipids) by column chromatography using an aminopropyl mini-column (Waters Sep-Pak® Cartridges; Waters Inc.; Milford, MA, USA). The FFA fraction was methylated with 20% BF3 in methanol (90 °C for 1 h), and the FAs were recovered as fatty acid methyl esters (FAMEs). The FAMEs were analyzed by gas chromatography (Agilent 5890 Series II; Agilent Technology Inc.; Santa Clara, CA, USA) equipped with a flame ionization detector. The FAMEs were identified and quantified based on retention times and a calibration curve using external standards (37 Component FAME Mix C4-C24; Supelco, Bellefonte, PA, USA). The peaks areas of each compound were integrated using ChromQuest 4.1 software, and total FA content ($\mu g \cdot mg^{-1}$) and composition (%) were determined.

2.6. Data Analysis and Statistics

Seasonal trends in the environmental parameters were analyzed by regression analysis using PASW Statistics (version 18.0; SPSS Inc.; Chicago, IL, USA). Univariate nonparametric (Kruskal-Wallis test and the Mann-Whitney *U* test) and correlation analyses were also performed using PASW. Multivariate analyses were performed using PRIMER version 6 [68]. We tested for differences in FA composition between the amphipod and red algal groups, and between the amphipod groups during different seasons using the analysis of similarity (ANOSIM). Similarities between the groups were further identified in a non-metric multi-dimensional scaling (MDS) ordination plot based on the Bray-Curtis similarity index. The relative contributions of the FAs to the similarity/dissimilarity identified in the MDS plot were analyzed using the similarity percentages program (SIMPER).

2.7. Supporting Data

Seawater temperature, as well as salinity and chlorophyll (Chl) data, were obtained from year-round monitoring at the station [69].

3. Results

3.1. Seasonal Variations in Salinity, Seawater Temperature, and Chl-a Concentrations

Salinity (psu) varied slightly (32.6 to 34.4), with a small decline seen during the summer months due to meltwater input (Figure 2a). In contrast, seawater temperature exhibited distinct variation (+1.97 to −1.87 °C) over the year, with the lowest monthly average occurring in September (−1.85 °C) and the highest in February (1.49 °C) (Figure 2b). The seawater was frozen (sea ice thickness up to 1.5 m, personal observation) for approximately 3 months from June 21 through October 2. During this period, the temperature was almost stable, remaining at the seawater freezing point (−1.8 °C).

Chl-*a* concentrations in the surface seawater peaked several times during summer (January–March) with the highest monthly average occurring in January (0.64 $\mu g \cdot L^{-1}$) (Figure 2c). The highest daily peak (1.28 $\mu g \cdot L^{-1}$) occurred on February 25. In addition, unusually late blooms occurred in May (monthly mean = 0.48 $\mu g \cdot L^{-1}$ with several daily peaks of ~1.05 $\mu g \cdot L^{-1}$). Thereafter, Chl-*a* concentrations declined sharply and remained at very low levels (<0.1 $\mu g \cdot L^{-1}$) during winter (June–August), with the lowest monthly average occurring in July (0.046 $\mu g \cdot L^{-1}$). The Chl-*a* values started to increase again in early

September, reaching levels similar to summer during late November (0.64–0.67 μg·L^{-1}). Overall, the variation in Chl-*a* closely followed the variation in seawater temperature (r = 0.66, n = 66, p < 0.001), while salinity was negatively correlated with temperature (r = −0.67, n = 266, p < 0.001).

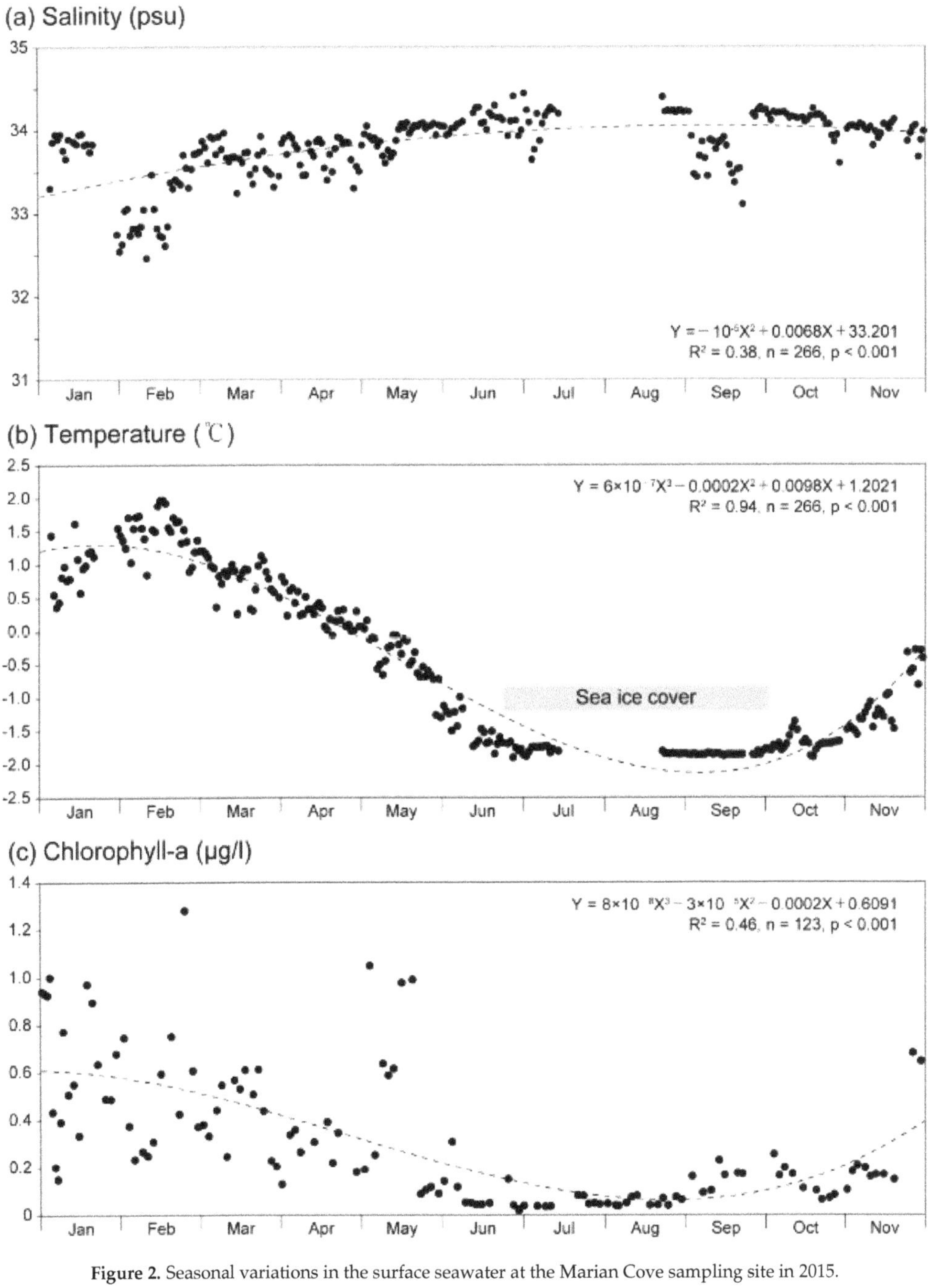

Figure 2. Seasonal variations in the surface seawater at the Marian Cove sampling site in 2015.

3.2. Seasonal Variations in the Isotopic Signatures of Gondogeneia antarctica and Its Potential Food Sources

Seasonal variations in $\delta^{13}C$ and $\delta^{15}N$ values for *G. antarctica* and SPOM are displayed in Figure 3a–d. The monthly average $\delta^{13}C$ and $\delta^{15}N$ values for *G. antarctica*, and its potential food sources during different months, are summarized in Table 1.

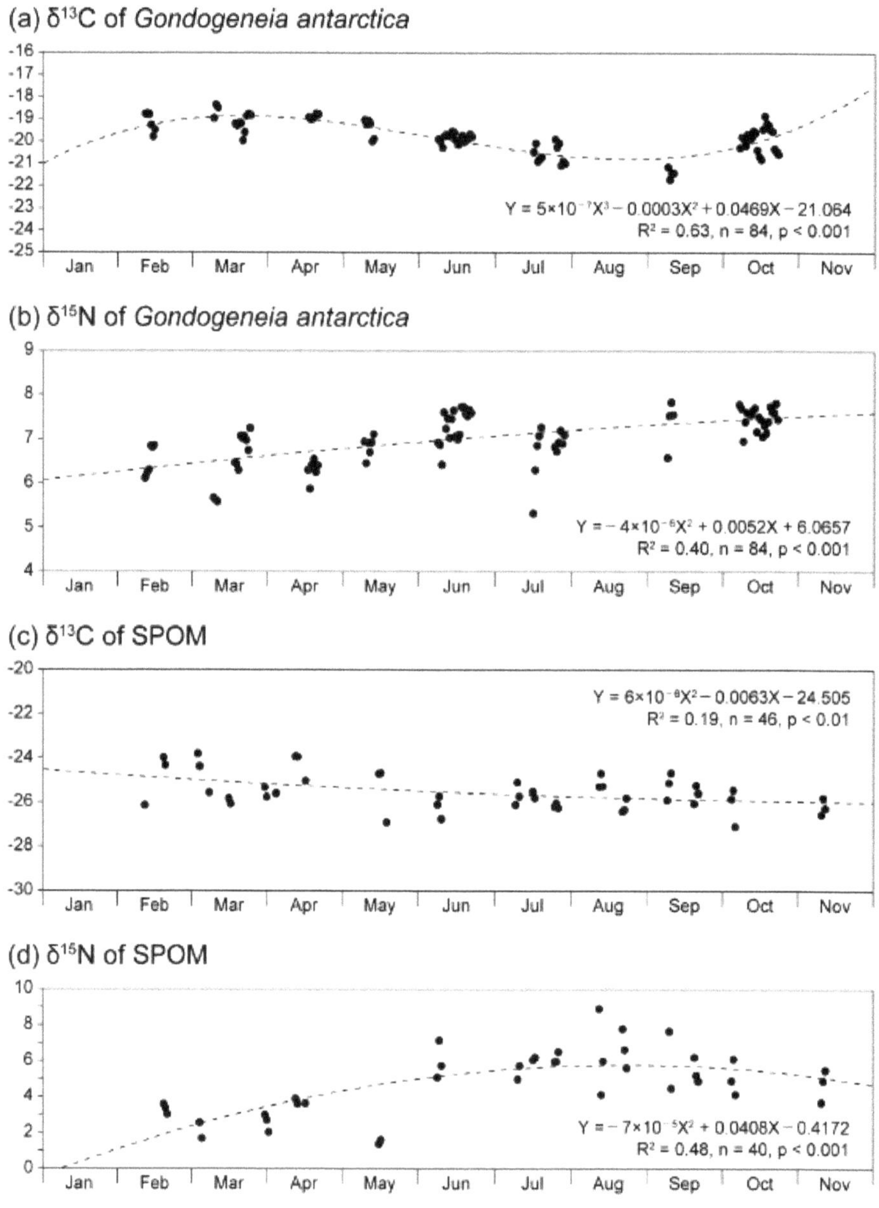

Figure 3. Seasonal variations in the $\delta^{13}C$ and $\delta^{15}N$ values for *Gondogeneia antarctica* (**a**,**b**) and suspended particulate organic matter (SPOM) (**c**,**d**) in the surface water of the sampling site in 2015.

Table 1. C and N stable isotopic ratios of the amphipod *Gondogeneia antarctica* and its potential food sources in various months. *n*, replicate numbers for the δ^{13}C and δ^{13}N analyses, except those indicated for SPOM δ^{15}N values. Numbers in parentheses for *G. antarctica* indicate the total number of individuals pooled for analysis. Mean and standard deviation (SD) values are presented. R: red algae; Br: brown algae. *n*: replicates of analysis. The data with * was cited from [30].

Specimens	Month	*n*	δ^{13}C (‰)		δ^{15}N (‰)		
			Mean	SD	Mean	SD	
Gondogeneia antarctica	Feb	6 (137)	−19.2	0.44	6.53	0.35	
	Mar	12 (174)	−19.1	0.45	6.52	0.61	
	Apr	6 (107)	−18.9	0.10	6.30	0.23	
	May	6 (151)	−19.4	0.42	6.85	0.23	
	Jun	18 (449)	−19.9	0.19	7.32	0.37	
	Jul	11 (94)	−20.6	0.42	6.78	0.55	
	Sep	4 (27)	−21.4	0.23	7.38	0.55	
	Oct	21 (598)	−19.9	0.53	7.49	0.24	
SPOM	Feb	3	−24.8	1.17	3.32	0.29	
	Mar	5	−25.1	0.99	2.12	0.61	
	Apr	6	−24.9	0.81	3.14	0.70	
	May	3	−25.4	1.29	1.48	0.16	(*n* = 2)
	Jun	3	−26.2	0.52	5.98	1.05	
	Jul	8	−25.8	0.40	5.91	0.48	(*n* = 7)
	Aug	6	−25.6	0.68	6.50	1.69	
	Sep	6	−25.4	0.52	5.68	1.28	(*n* = 5)
	Oct	3	−26.1	0.85	5.04	1.00	
	Nov	3	−26.2	0.39	4.69	0.91	
Macroalgae							
Iridaea cordata (R)	Mar	2	−19.4	2.69	5.21	0.26	
	May	4	−23.1	2.00	5.03	0.39	
Palmaria decipiens (R)	May	1	−19.4		5.12		
	Oct	2	−22.3	2.65	4.18	0.06	
	Jan *	2	−23.1	0.94	3.33	0.75	
Gigartina skottsbergii (R)	May	1	−21.1		4.90		
	Oct	1	−23.4		3.55		
	Jan *	2	−23.1	1.32	3.26	0.36	
Curdiea racovitzae (R)	May	1	−17.4		5.87		
	Jan *	2	−14.2	0.63	4.46	0.44	
Phaeurus antarcticus (Br)	May	2	−32.0	1.22	2.78	0.06	
Desmarestia sp. (Br)	Oct	1	−27.3		3.45		
Benthic diatom	Jan *	3	−23.3	0.26	3.55	0.30	

The *G. antarctica* δ^{13}C values (monthly averages: max = −18.9‰ in April, min = −21.4‰ in September) were positively correlated with seawater temperature (Pearson's correlation coefficient, r = 0.67, n = 84, $p < 0.001$) and the Chl-*a* concentration (r = 0.62, n = 84, $p < 0.001$). The overall variations were small but showed more significant enrichment (Mann–Whitney *U*-test, $p < 0.001$) during the summer–early fall period (February–April) (−19.1 ± 0.4‰, n = 24) than during the fall–winter period (May–September) (−20.2 ± 0.7‰, n = 39). In contrast, the *G. antarctica* δ^{15}N value (6.3‰ in April to 7.5‰ in October) was negatively correlated with seawater temperature (r = −0.60, n = 84, $p < 0.001$) and Chl-*a* concentrations (r = −0.47, n = 84, $p < 0.001$), with significant enrichment (p < 0001) seen in the winter–spring period (May–October) (7.2 ± 0.5‰, n = 60) compared to the summer–fall period (February–April) (6.5 ± 0.4‰, n = 24).

The seasonal variation patterns in SPOM δ^{13}C and δ^{15}N (Figure 3c,d) were similar to those of *G. antarctica*. The SPOM δ^{13}C showed a small but significant variation over the year (monthly means = −24.8‰ to −26.2‰) and was positively correlated with seawater temperature (Pearson's correlation coefficient (r) = 0.43, n = 46, $p < 0.01$) and Chl-*a* concentrations (r = 0.36, n = 46, $p < 0.05$). In contrast, the SPOM δ^{15}N values (1.5‰ to 6.5‰) were negatively correlated with seawater temperature (r = −0.74, n = 40, p <

0.001) and Chl-*a* concentrations (r = −0.78, n = 40, p < 0.001). The SPOM δ^{15}N value showed a relatively large (p < 0.001) increase in the winter season (June–September) (6.1 ± 0.9‰, n = 21) compared to the summer–fall period (February–April) (3.0 ± 0.7‰, n = 11). The SPOM δ^{15}N value sharply declined in May, during which unusual Chl-*a* peaks occurred.

Isotopic values of macroalgae were compared only for the two contrasting seasons (May vs. January, March, and October), as sampling was impossible during the winter season (June–September), due to the formation of sea ice (pack ice, fast ice) in the sampling site. The macroalgae δ^{13}C values varied widely among species, particularly between red and brown algae. Within the red algal species, *C. racovitzae* (−15.6 to −17.4‰,) was much more enriched with δ^{13}C (p < 0.001) than the other three species (*I. cordata* −21.3 to −22.3‰, *P. decipiens* −19.4 to −20.5‰, and *G. skottsbergii* −21.1 to −23.4‰). The brown algae *Desmarestia* sp. and *P. antarcticus* were most depleted with δ^{13}C and δ^{15}N (*Desmarestia* sp. −27.3‰ and 3.5‰, *P. antarcticus* −32.0‰ and 2.8‰). The macroalgae δ^{13}C varied little seasonally, despite the wide variation among species. On the other hand, the macroalgae δ^{15}N values were significantly (Mann-Whitney *U*-test, p = 0.008) enriched in May (5.1 ± 0.4‰, n = 7) compared to the other months (January, March, and October) (4.0 ± 0.8‰, n = 11).

3.3. Isotopic Signatures of Gondogeneia antarctica and Its Potential Food Sources

The δ^{13}C and δ^{15}N signatures of *G. antarctica* and its potential carbon sources are displayed in a dual plot (Figure 4). The amphipod and groups of potential food sources (SPOM, red algae, brown algae, and benthic diatoms) during different months were well distinguished by their δ^{13}C and δ^{15}N values. The monthly *G. antarctica* δ^{13}C values ranged from −18.9 to −21.4‰, which were close to those of three red algal species (−19.4 to −23.4‰) and benthic diatom (−23.3‰). On the other hand, the SPOM (monthly means = −24.8‰ to −26.2‰) and the brown algae (*Desmarestia* sp. −27.3‰ and *P. antarcticus* −32.0‰) exhibited far more depleted δ^{13}C values.

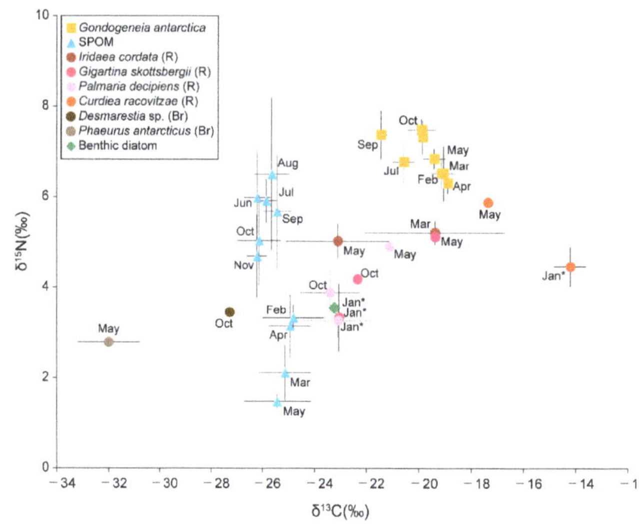

Figure 4. Seasonal variations in the δ^{13}C and δ^{15}N values of *Gondogeneia antarctica* and its potential food sources (suspended particulate organic matter, SPOM; red algae, R, and brown algae, Br). Sampling was conducted from February to November 2015. The data for benthic diatom and additional data for macroalgae (* marked) were obtained from a previous study performed in the study area [30]. Mean ± standard deviation values are presented. Refer to Table 1 for details.

3.4. Total FA Content and Composition in Gondogeneia antarctica and Red Algae

The FA contents and composition of *G. antarctica* and its presumptive major diet, red algae are shown in Table 2. The most dominant FAs in the amphipod throughout the year were C16:0 (mean = 26.2% in June, 33.5% in March), followed by C20:5$_{(n\text{-}3)}$ (14.6%–23.9%), C18:2$_{(n\text{-}6)}$ (10.6%–15.3%), C18:1$_{(n\text{-}9)}$ (7.0%–11.8%), C16:1$_{(n\text{-}9)}$ (4.6%–10.3%), and C22:6$_{(n\text{-}3)}$ (3.5%–5.5%).

The total FA contents in *G. antarctica* continued to increase from mid-summer (February) until reaching a peak in May (60.5 ± 22.1 μg·mg^{-1}dw, n = 6). The total remained at similar levels during the winter season and then declined to its lowest level in October (23.2 ± 7.8 μg·mg^{-1}dw, n = 13) (Figure 5a). Among the major FAs, C16:0 and C20:5$_{(n\text{-}3)}$ showed relatively large variations. Of note, C20:5$_{(n\text{-}3)}$ distinctly increased in the winter period in terms of both the absolute and relative composition. (Figure 5b,c). C16:0 and C18:1$_{(n\text{-}9)}$, C20:5$_{(n\text{-}3)}$, and C20:4$_{(n\text{-}6)}$ were the major FAs in the red algae analyzed in this study. However, the FA composition varied among algal species. C16:0 and C18:1$_{(n\text{-}9)}$ were most abundant in *I. cordata* and *P. decipiens*, while C16:0 and C20:5$_{(n\text{-}3)}$ were detected as the most dominant in *G. skottsbergii* and *C. racovitzae*, respectively.

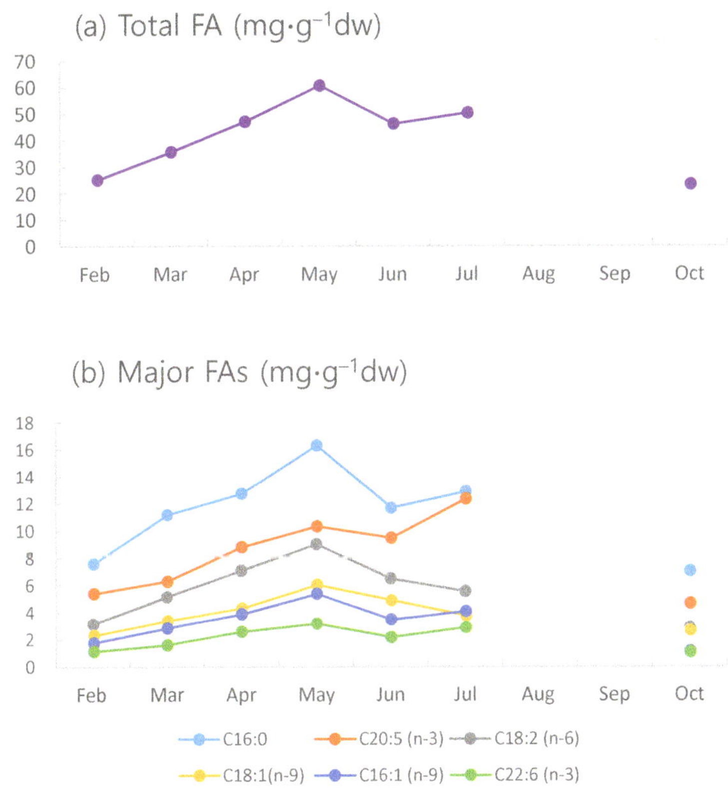

Figure 5. *Cont.*

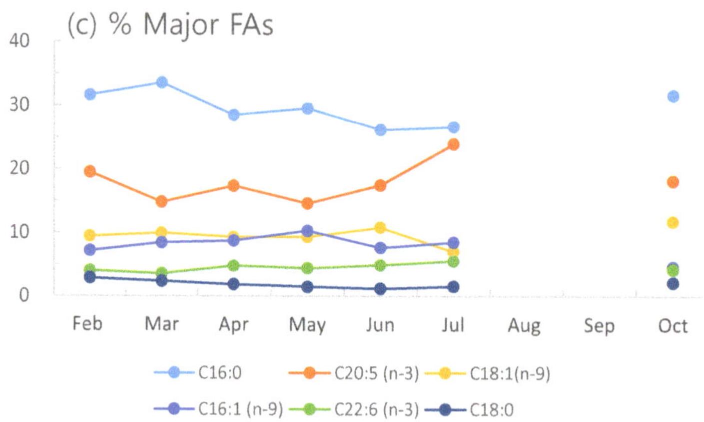

Figure 5. Seasonal variations in the *Gondogeneia antarctica* fatty acids (FAs). (**a**) Total contents; (**b**) absolute concentration of major FAs; (**c**) % composition of the major FAs. Mean values are presented. Refer to Table 2 for details.

Nonparametric MDS analysis (Figure 6) revealed that a small but distinct shift in *G. antarctica* FA composition occurred from spring–summer (February, October) to fall–winter (March, April, June, and July) (ANOSIM test: Global R = 0.17, p = 0.014). The MDS plot also showed that the spring–summer composition was closer to the red algal FA composition. The SIMPER analysis (Table 3) showed that the dissimilarity between the amphipod and red algae FA compositions decreased slightly from 42% during fall–winter to 36% during spring–summer.

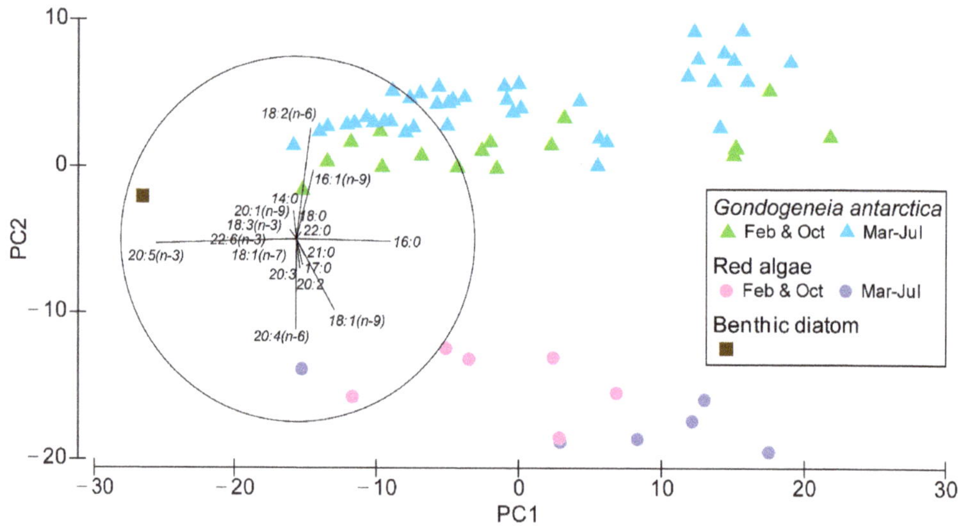

Figure 6. Non-parametric multidimensional scaling (MDS) based on the Bray–Curtis similarity matrix for the fatty acid (FA) compositions of *Gondogeneia antarctica* and red algae during the two contrasting seasons. Small but distinct differences in the FA composition were detected in *G. antarctica* between the spring–summer (February and October) and fall–winter (March–July). The FA composition of *G. antarctica* was closer to the red algal FA composition during spring–summer than early winter (March–May). Benthic diatom data are from [13].

Table 2. Total fatty acid (FA) content and % composition in *Gondogeneia antarctica* and its major diet, namely four red algal species (*Iridaea cordata, Palmaria decipiens, Curdiea racovitzae*, and *Gigartina skottsbergii*). The figures in bold indicate major FAs. FAs with mean values <0.5% are not shown in the table. (-) FA not detected. Mean and standard deviation (SD) values are presented.

	Gondogeneia antarctica							Red Algae			
	Feb (n = 4)	Mar (n = 3)	Apr (n = 6)	May (n = 6)	Jun (n = 22)	Jul (n = 3)	Oct (n = 13)	*Iridaea cordata* (n = 4)	*Palmaria decipiens* (n = 3)	*Gigartina skottsbergii* (n = 3)	*Curdiea racovitzae* (n = 2)
Total free FA (mg·g^{-1} dw)	25.2 ± 6.59	35.6 ± 14.7	47.0 ± 14.3	60.5 ± 22.1	46.2 ± 19.2	50.2 ± 12.1	23.2 ± 7.77	60.6 ± 10.5	59.9 ± 4.54	63.2 ± 1.78	46.5 ± 1.14
% of FAs											
C14:0	1.49 ± 0.21	2.85 ± 0.71	3.29 ± 0.66	4.89 ± 1.01	3.27 ± 0.73	3.2 ± 0.73	1.5 ± 0.44	-	0.22 ± 0.38	-	0.28 ± 0.4
C16:1 (n-9)	**7.17** ± 1.45	**8.41** ± 2.67	**8.72** ± 2.29	**10.3** ± 4.76	**7.64** ± 1.68	**8.48** ± 2.51	**4.61** ± 1	0.95 ± 1.09	0.62 ± 1.07	0.24 ± 0.05	1.99 ± 2.82
C16:0	**31.6** ± 8.15	**33.6** ± 7.93	**28.5** ± 6.07	**29.5** ± 8.68	**26.2** ± 4.53	**26.7** ± 6.59	**31.7** ± 7.3	**31.8** ± 4.55	**26.2** ± 2.3	**42.7** ± 2.7	**20.0** ± 1.7
C17:0	1.08 ± 0.25	1.03 ± 0.27	0.6 ± 0.08	0.7 ± 0.08	0.88 ± 0.23	0.67 ± 0.25	1.33 ± 0.36	4.21 ± 1.64	5.83 ± 5.05	0.56 ± 0.04	4.72 ± 4.15
C18:3 (n-3)	1.84 ± 1.02	1.45 ± 0.79	2.8 ± 1.55	3.06 ± 1.4	2.7 ± 1.25	4.07 ± 0.4	1.05 ± 0.69	1.52 ± 0.45	2.07 ± 0.45	1.21 ± 0.25	0.52 ± 0.06
C18:2 (n-6)	**12.9** ± 2.05	**14.9** ± 2.05	**15.3** ± 3.39	**14.1** ± 4.14	**14.2** ± 2.25	**10.6** ± 2.95	**12.6** ± 2.01	2.21 ± 1.12	1.81 ± 1.16	0.81 ± 0.07	2.72 ± 1.72
C18:1(n-9)	**9.42** ± 1.56	**9.87** ± 1.54	**9.26** ± 1.76	**9.3** ± 2.97	**10.8** ± 1.71	**7** ± 2.97	**11.8** ± 2.25	**24.9** ± 3.64	**21.1** ± 2.19	**9.31** ± 0.7	**12.5** ± 0.51
C18:1(n-7)	0.53 ± 0.3	0.47 ± 0.1	0.45 ± 0.13	1.25 ± 1.03	0.56 ± 0.3	0.25 ± 0.23	1.08 ± 2.96	1.11 ± 0.48	0.3 ± 0.51	0.22 ± 0.04	1.09 ± 1.22
C18:0	2.84 ± 0.58	2.31 ± 0.29	1.84 ± 0.29	1.44 ± 0.73	1.19 ± 0.62	1.5 ± 0.34	2.12 ± 0.7	-	0.47 ± 0.82	-	-
C20:5 (n-3)	**19.4** ± 10.4	**14.8** ± 10.4	**17.3** ± 8.2	**14.6** ± 8.57	**17.4** ± 9.33	**23.9** ± 3.86	**18.2** ± 9.33	**6.14** ± 1.52	**11.8** ± 1.39	**27.3** ± 3.38	**27.8** ± 3.91
C20:4 (n-6)	1.05 ± 0.31	0.67 ± 0.17	0.58 ± 0.21	0.57 ± 0.18	1.27 ± 0.44	1.46 ± 0.03	1.61 ± 0.45	**10.0** ± 1.46	**11.5** ± 2.85	**11.4** ± 1.51	**14.9** ± 0.86
C20:3	0.94 ± 0.49	0.77 ± 0.39	1.29 ± 0.5	1.23 ± 0.29	1.98 ± 1.96	1.63 ± 0.08	0.92 ± 0.42	1.45 ± 0.32	4.33 ± 1.57	0.41 ± 0.03	0.46 ± 0.29
C20:2	0.6 ± 0.05	0.56 ± 0.05	0.42 ± 0.04	0.43 ± 0.04	0.54 ± 0.26	0.73 ± 0.13	0.77 ± 0.24	4.43 ± 1.35	5.56 ± 3.48	1.36 ± 0.1	3.45 ± 1.41
C20:1 (n-9)	1.3 ± 0.12	1.73 ± 0.44	2.63 ± 0.94	1.57 ± 0.98	1.58 ± 0.75	1.77 ± 0.4	1.44 ± 0.4	0.26 ± 0.08	0.77 ± 0.25	0.86 ± 0.08	1.49 ± 0.83
C21:0	-	-	-	0.05 ± 0.11	1.56 ± 3.56	-	0.16 ± 0.44	3.95 ± 3.35	0.59 ± 0.56	1.32 ± 0.13	2.53 ± 0.89
C22:6 (n-3)	3.99 ± 2.65	3.52 ± 3.47	4.74 ± 3.01	4.36 ± 2.96	4.36 ± 2.36	5.5 ± 1.37	4.19 ± 1.82	4.41 ± 1.41	4.9 ± 2.17	0.82 ± 0.07	4.08 ± 1.4
C22:0	1.78 ± 0.59	1.38 ± 0.46	0.9 ± 0.24	0.78 ± 0.53	0.8 ± 0.55	0.99 ± 0.39	1.88 ± 0.67	-	-	-	-
C23:0	0.97 ± 0.32	0.74 ± 0.25	0.49 ± 0.15	0.41 ± 0.28	0.4 ± 0.24	0.54 ± 0.21	0.97 ± 0.42	-	-	-	-
sum(%)	98.9	99.0	99.1	98.5	98.0	98.9	97.8	97.4	98.0	98.5	98.5

Table 3. Results of SIMPER analyses of dissimilarities in the fatty acid (FA) composition between groups. Average dissimilarity (%) and the most discriminant FAs are presented. Spring–summer (February and October), winter (March–July). * data for benthic diatoms are from [13].

Group	Dissimilarity (%)	Most Discriminant FAs (% Contribution)
Within *Gondogeneia antarctica* Spring-summer vs. fall-winter	21.6	20:5(25)/16:0(20)/16:1(8.3)/18:2(7.0)/18:1(6.6)/22:6(6.3)
Gondogeneia antarctica vs. red algae		
Spring-summer	35.8	18:2(16)/20:4(15)/16:0(15)/20:5(14)/18:1(15)/16:1(5.9)
Fall-winter	42.1	18:1(15)/18:2(14)/20:5(14)/20:4(13)/16:1(9.3)/16:0(7.9)
Gondogeneia antarctica vs. benthic diatom *		
Spring-summer	57.3	16:0(26)/20:5(16)/18:2(13)/18:1(12)/14:0(12)/16:1(5.8)
Fall-winter	56.4	16:0(21)/20:5(18)/18:2(15)/18:1(10)/14:0(9.6)/16:1(9.4)

The SIMPER analysis showed that the FAs contributing most to the dissimilarities in *G. antarctica* between the two contrasting seasons were $C20:5_{(n\text{-}3)}$ (25%) and C16:0 (20%), followed by $C16:1_{(n\text{-}9)}$ (8.3%), $C18:2_{(n\text{-}6)}$ (7.0%), $C18:1_{(n\text{-}9)}$ (6.6%), and $C22:6_{(n\text{-}3)}$ (6.3%). Of note, $C20:5_{(n\text{-}3)}$ increased to its highest value in July, in terms of both the absolute and relative compositions, while the other FAs decreased or changed slightly during the same period. The MDS plot showed that the FA composition of red algae was much closer to that of *G. antarctica* than the composition of benthic diatoms. The FA composition of benthic diatoms was distinct from *G. antarctica*.

4. Discussion

4.1. Isotopic Signatures and Seasonal Dietary Shift of G. antarctica

The isotopic dual plot (Figure 4) indicated a higher contribution of red algae to *G. antarctica* diet relative to the other sources investigated in the study. The *G. antarctica* $\delta^{13}C$ values (−18.9 to −21.4‰) were close to the values of the 3 red algae (*I. cordata* −21.3 to −22.3‰, *P. decipiens* −19.4 to −20.5‰, and *G. skottsbergii* −21.1 to −23.4‰) and benthic diatoms (−23.3‰), while the values were much heavier than those of SPOM (−24.8‰ to −26.2‰) and brown algae (*Desmarestia* sp. −27.3‰ and *P. antarcticus* −32.0‰). It seemed that the red algae apparently contributed most to the *G. antarctica* diet throughout the investigation period, considering a fractionation factor of 0.8‰ in consumer $\delta^{13}C$ relative to its diet [59,70]. The red algae are known to be palatable to many herbivores, detritivores and omnivores [10,34,54–56]. In particular, *P. decipiens* is the most preferred by *G. antarctica* [10,56,71], and is one of the most common macroalgae in shallow waters adjacent to this study area and other similar environments around King George Island [22,53,72]. As reflected in the isotopic signatures, benthic diatoms seemed to partially contribute to the *G. antarctica* diet, while contribution of SPOM appears to be relatively small.

The *G. antarctica* $\delta^{13}C$ and $\delta^{15}N$ values varied distinctly with seasonal variations in seawater temperature, salinity, and Chl-*a* concentrations. However, the overall variation was small, indicating that the main diets of the amphipod, mainly red algae and partly benthic diatoms, did not change much throughout the year. The slight enrichment of $\delta^{13}C$ in *G. antarctica* during the summer season (February–April) was likely due to increased availability of a fresh red algal diet. Around King George Island and the adjacent islands, the highest growth rates of these algal species were attained in austral spring (October–November) (*P. decipiens*) or summer (December) (*I. cordata* and *G. skottsbergii*), when the light conditions were most favorable for photosynthetic activities [24,73–75].

The *G. antarctica* isotopic values in this study were similar to those reported for the same species and other amphipod species in the WAP region [13,59]. Like other macroalgal grazers, the *G. antarctica* $\delta^{13}C$ are heavier than the values of herbivorous suspension feeders whose diets are derived mainly from phytoplankton [76], and this is consistent with the result of this study that red algae are a primary carbon source in the *G. antarctica* diet.

Thus, the results of this study revealed that the preference of *G. antarctica* for macroalgal diet persisted throughout much of the year. However, the *G. antarctica* $\delta^{15}N$ values maintained rather high (6.3~7.5‰) as for primary consumers, suggesting that dietary contribution from other food sources (e.g., crustacean parts and sponge spicules) [9,10,13,34] should be considered in future studies.

4.2. Contribution of Detrital Matter to the *Gondogeneia antarctica* Diet during the Fall-Winter

Our results also suggest that degrading organic matter is an important food source for *G. antarctica*, particularly under subdued primary production during fall and winter. The $\delta^{15}N$ values increased significantly during the winter period in *G. antarctica* and its presumptive main diet, red algae, and also in macroalgae and SPOM (Figure 3). In particular, the SPOM $\delta^{15}N$ reached the peak values (6.1 ± 0.9‰, n = 21) in the winter season [June–September], which were two times higher than the values in the summer-early fall period [February–April] (3.0 ± 0.7‰, n = 11). This observation indicated that, during winter, a significant portion of the diet of *G. antarctica* is in the form of detritus, which was enriched with $\delta^{15}N$ through colonization and degradation by microbes, and sometimes also by protozoans and meiofauna [77–79].

After the growth peaks in spring-summer, macroalgal growth decrease during the fall and winter seasons around King George Island and the adjacent islands, particularly under the winter sea-ice cover, when light conditions for photosynthetic activities are unfavorable [24,73–75]. Therefore, fresh algal material is likely restricted in availability during fall-winter. *G. antarctica* is known to feed year-round [32,33], and there may be other alternative food sources for *G. antarctica* during this period. Macroalgal fragments seem to create a food bank for benthic communities in shallow Antarctic waters, as degradation processes by microbes may be slow due to low temperature [79,80]. Several studies have reported that a significant amount of carbon derived from macroalgal detritus enters the benthic food web in nearshore waters of the WAP [9,59,81–84].

Erosion and degradation of benthic macroalgae likely occur through several processes, such as mechanical abrasion by floating icebergs, tides, waves and storms, sea ice formation, and grazing [81,84]. Grazing by herbivores, such as amphipods, also promotes microbial degradation [79]. Almost all of the brown algae are unpalatable to grazers due to their chemical defenses and/or stiff texture [34,71]. However, brown algae can be palatable in detrital form [85].

Degrading macroalgal biomass seems to reach a peak in fall–early winter in King George Island, and to be consumed as a main diet by shallow-water herbivores like *G. antarctica*. In the adjacent Admiralty Bay, a large amount of decaying macroalgae were present on the shore in late April–early June, and the peak abundance and biomass of *G. antarctica* occurred during the same period [9]. In this study, we also observed a large accumulation of decaying macroalgae on the shore in April–May right before the sea water started to freeze (Figure 7). This observation supports the isotopic results; a significant portion of the diet of *G. antarctica* is in the form of detritus during the fall and winter periods. Benthic diatoms are also likely preserved as a food bank after summer blooms and utilized as alternative food sources to support benthic communities during winter [80,86]. In addition, sea ice may provide a habit for growth of epontic algae (mostly diatoms), which are grazed by herbivores like amphipods. Several studies reported that dense populations of adult and juvenile *G. antarctica* [33] and other amphipod species [32] were associated with micro-algal blooms under sea ice during mid-winter in the WAP. In the study area, the sea surface was frozen for about 3 months (late June through September) during the investigation period (Figure 2).

Figure 7. Macroalgal debris amassed on the shore of Marian Cove from April to early May 2015: (**a**) Mostly *Desmarestia* spp., (**b**) red algal fragments, (**c**) *Iridaea cordata,* (**d**) *Gigartina skottsbergii*, and (**e**) *Curdiea racovitzae*.

4.3. FA Composition in Gondogeneia antarctica and Its Major Food Sources

The FA results supported the isotopic results that the benthic primary producers, preferably red algae, contributed substantially to the *G. antarctica* diet throughout much of the year. The FA composition of *G. antarctica* showed small but distinct seasonal variation, particularly between the spring–summer and the fall–winter periods (Table 2, Figures 5 and 6). $C20{:}5_{(n\text{-}3)}$ and C16:0 collectively accounted for most (45%) of the dissimilarity between the two contrasting periods (Table 3). The red algal FA composition was more similar to that of *G. antarctica* in spring–summer than winter (Figure 6, Table 3), indicating that fresh red algae were more available to the amphipod during this period.

Notably, $C20{:}5_{(n\text{-}3)}$ in *G. antarctica* increased to its highest value in July in terms of both the absolute and relative compositions, while the other FAs decreased or changed only slightly during the same period. The increase of $C20{:}5_{(n\text{-}3)}$ in winter may be related, in part, to the increased contribution of benthic diatoms to the *G. antarctica* diet. A previous study reported that the gut contents and FA composition of *G. antarctica* reveal an important contribution of benthic diatoms [13]. It is well known that $C20{:}5_{(n\text{-}3)}$ occurs as a major FA in marine benthic diatoms and red algae [13,87], and has been used as a reliable dietary tracer in various marine benthic habitats from tropical to polar regions [88].

The FA composition of *G. antarctica* was similar to those of most Antarctic amphipods. Among the major FAs in *G. antarctica*, C16:0, $C20{:}5_{(n\text{-}3)}$, $C18{:}1_{(n\text{-}9)}$, and $C22{:}6_{(n\text{-}3)}$ occur as major FAs in most amphipods in the peninsular region [13,89], and C16:0, $C20{:}5_{(n\text{-}3)}$, and $C18{:}1_{(n\text{-}9)}$ were also the major FAs in the red algae analyzed in this and other studies [13]. In particular, $C20{:}5_{(n\text{-}3)}$ and C16:0 predominated, constituting most (41%–57%) of the FAs in this amphipod throughout the year.

Total FA content varied distinctly, with a peak seen in winter (Figure 5a). However, this pattern was unlikely to be related to variations in the diet, but rather to the reproductive cycle of *G. antarctica*, which is known to have two spawning peaks (at the end of February and at the end of September) [57]. We also observed some brooding individuals in July and early October samples, supporting the idea that reproductive development proceeded during the winter, contributing to the increase in total FA content.

4.4. Prospects for Climate Changes

The WAP is rapidly warming and deglaciating under climate change [5]. The glacial retreats are opening up new habitats for marine benthic organisms [90], and sublittoral rocky substrates are increasingly colonized by macroalgae [91], seemingly providing more food for *G. antarctica* and other herbivores. In MC, *Desmarestia* spp. and the red alga *Plocamium cartilagineum*, species unpalatable to most herbivores, have increased significantly during the last 30 years, while *P. decipiens*, a species most preferred by *G. antarctica*, has decreased [53]. In addition, a recent study indicated that climate change could alter the biochemical composition of macroalgae, but not the feeding preference of *G. antarctica* [92]. Given that only a few macroalgal species are palatable to herbivores like *G. antarctica*, the decrease in the palatable algal population could affect the survival of this species, alter food-web structures, and eventually lead to ecosystem changes.

5. Conclusions

The SI and FA results in this study revealed that *Gondogeneia antarctica* rely substantially on food sources derived from benthic primary producers, mostly red algae and their detrital forms and partly benthic diatoms. However, the relatively high $\delta^{15}N$ values in *G. antarctica*, as for primary consumers, indicated additional dietary sources such as animal parts. The isotopic signatures varied with only small shifts seen throughout the year, indicating that contribution from the main dietary sources did not change much throughout the year. The overall results suggest that *G. antarctica* is not much affected by the extreme seasonality of pelagic primary production, and this strategy seems to have allowed *G. antarctica* to adapt successfully to shallow Antarctic nearshore waters. However, ongoing climate change is altering the distribution and biochemical composition of macroalgae favored by the amphipod, which could affect its adaptation and survival.

Author Contributions: I.-Y.A.: conceptualization, sample collection, data curation, writing of the original draft and editing, and project administration; F.E.-P.: conceptualization, methodology (sample analysis), data curation, and writing of the original draft; S.-Y.H.: methodology (sample analysis), data curation, and writing–review and editing; S.R.: writing–review and editing; D.-U.K.: data curation and graphics. All authors have read and agreed to the published version of the manuscript.

Funding: This research was funded by the Korea Polar Research Institutes [grant nos. PE17070, PE 20120, and PE21110].

Acknowledgments: We thank Ji Hee Kim for identifying the macroalgal samples and Jong-Ku Gal for reviewing the manuscript.

Conflicts of Interest: The authors declare no conflict of interest.

References

1. Ducklow, H.W.; Fraser, W.R.; Meredith, M.P.; Stammerjohn, S.E.; Doney, S.C.; Martinson, D.G.; Sailley, S.F.; Schofield, O.M.; Steinberg, D.K.; Venables, H.J.; et al. West Antarctic Peninsula: An ice-dependent coastal marine ecosystem in transition. *Oceanography* **2013**, *26*, 190–203. [CrossRef]
2. Constable, A.J.; Melbourne-Thomas, J.; Corney, S.P.; Arrigo, K.R.; Barbraud, C.; Barnes, D.K.A.; Bindoff, N.; Boyd, P.; Brandt, A.; Costa, D.; et al. Climate change and Southern Ocean ecosystems I: How changes in physical habitats directly affect marine biota. *Glob. Chang. Biol.* **2014**, *20*, 3004–3025. [CrossRef] [PubMed]
3. Cauvy-Fraunie, S.; Dangles, O. A global synthesis of biodiversity responses to glacier retreat. *Nature Ecol. Evol.* **2019**, *3*, 1675–1685. [CrossRef]
4. Gutt, J.; Isla, E.; Xavier, J.C.; Adams, B.J.; Ahn, I.; Cheng, C.C.; Colesie, C.; Cummings, V.J.; di Prisco, G.; Griffiths, H.; et al. Antarctic ecosystems in transition—Life between stresses and opportunities. *Boil. Rev.* **2021**, *96*, 798–821. [CrossRef] [PubMed]
5. Cook, A.J.; Vaughan, D.G.; Luckman, A.J.; Murray, T. A new Antarctic Peninsula glacier basin inventory and observed area changes since the 1940s. *Antarct. Sci.* **2014**, *26*, 614–624. [CrossRef]
6. Smale, D.A.; Brown, K.M.; Barnes, D.K.; Fraser, K.P.; Clarke, A. Ice scour disturbance in Antarctic waters. *Science* **2008**, *321*, 371. [CrossRef]
7. Barnes, D.K.A.; Souster, T. Reduced survival of Antarctic benthos linked to climate-induced iceberg scouring. *Nat. Clim. Chang.* **2011**, *1*, 365–368. [CrossRef]

8. Barnes, D.K.A. Iceberg killing fields limit huge potential for benthic blue carbon in Antarctic shallows. *Glob. Chang. Biol.* **2017**, *23*, 2649–2659. [CrossRef]
9. Jażdżewski, K.; De Broyer, C.; Pudlarz, M.; Zielinski, D. Seasonal fluctuations of vagile benthos in the uppermost sublittoral of a maritime Antarctic fjord. *Polar Biol.* **2001**, *24*, 910–917. [CrossRef]
10. Huang, Y.M.; McClintock, J.B.; Amsler, C.D.; Peters, K.J.; Baker, B.J. Feeding rates of common Antarctic gammarid amphipods on ecologically important sympatric macroalgae. *J. Exp. Mar. Biol. Ecol.* **2006**, *329*, 55–65. [CrossRef]
11. Huang, Y.M.; Amsler, M.O.; McClintock, J.B.; Amsler, C.D.; Baker, B.J. Patterns of gammaridean amphipod abundance and species composition associated with dominant subtidal macroalgae from the western Antarctic Peninsula. *Polar Biol.* **2007**, *30*, 1417–1430. [CrossRef]
12. Aumack, C.F.; Amsler, C.D.; Mcclintock, J.B.; Baker, B.J. Changes in amphipod densities among macroalgal habitats in day versus night collections along the Western Antarctic Peninsula. *Mar. Biol.* **2011**, *138*, 1879–1885. [CrossRef]
13. Aumack, C.F.; Lowe, A.T.; Amsler, C.D.; Amsler, M.O.; McClintock, J.B.; Baker, B.J. Gut content, fatty acid, and stable isotope analyses reveal dietary sources of macroalgal-associated amphipods along the western Antarctic Peninsula. *Polar Biol.* **2017**, *40*, 1371–1384. [CrossRef]
14. Jażdżewska, A.M.; Siciński, J. Assemblages and habitat preferences of soft bottom Antarctic amphipoda: Admiralty Bay case study. *Polar Biol.* **2017**, *40*, 1845–1869. [CrossRef]
15. Elias-Piera, F.; Rossi, S.; Petti, M.A.V.; Campos, L.S.; Valério-Berardo, M.T.; Corbisier, T.N. Faunal associated with morphologically distinct macoralgae from Admiralty Bay, King George Island (Antarctica). *Polar Biol.* **2020**, *43*, 1535–1547. [CrossRef]
16. De Broyer, C.; Lowry, J.K.; Jażdżewski, K.; Robert, H. Catalogue of the Gammaridean and Corophiidean Amphipoda (Crustacea) of the Southern Ocean with distribution and ecological data. *Bull. Inst. R Sci. Nat. Belg. Biol.* **2007**, *77*, 1–325.
17. Jażdżewski, K.; Teodorczyk, W.; De Broyer, C.; Konopacka, A. Survey and distributional patterns of the amphipod fauna of Admiralty Bay, King George Island, South Shetland Islands. *Pol. Polar Res.* **1991**, *12*, 461–472.
18. Jażdżewski, K.; Teodorczyk, W.; Siciński, J.; Kontek, B. Amphipod crustaceans as an important component of zoobenthos of the shallow Antarctic sublittoral. *Hydrobiologia* **1991**, *223*, 105–117. [CrossRef]
19. Dauby, P.; Scailteur, Y.; De Broyer, C. Trophic diversity within the eastern Weddell Sea amphipod community. *Hydrobiologia* **2001**, *443*, 69–86. [CrossRef]
20. Nyssen, F.; Brey, T.; Lepoint, G.; Bouquegneau, J.-M.; De Broyer, C.; Dauby, P. A stable isotope approach to the eastern Weddell Sea trophic web: Focus on benthic amphipods. *Polar Biol.* **2002**, *25*, 280–287. [CrossRef]
21. Nyssen, F.; Brey, T.; Dauby, P.; Graeve, M. Trophic position of Antarctic amphipods—Enhanced analysis by a 2-dimensional biomarker assay. *Mar. Ecol. Prog. Ser.* **2005**, *300*, 135–145. [CrossRef]
22. Oliveira, E.C.; Absher, T.M.; Pellizzari, F.M.; Oliveira, M.C. The seaweed flora of Admiralty Bay, King George Island, Antarctic. *Polar Biol.* **2009**, *32*, 1639–1647. [CrossRef]
23. Jażdżewska, A.M. Soft bottom sublittoral amphipod fauna of Admiralty Bay, King George Island, Antarctic. *Oceanol. Hydrobiol. Stud.* **2011**, *40*, 1–10. [CrossRef]
24. Wiencke, C.; Amsler, C.D. Chapter 13 Seaweeds and Their Communities in Polar Regions. In *Seaweed Biology*; Wiencke, D., Bischof, K., Eds.; Springer: Berlin/Heidelberg, Germany, 2012; pp. 265–291.
25. Kim, J.-H.; Jażdżewska, A.; Choi, H.-G.; Kim, W. The first report on Amphipoda from Marian Cove, King George Island, Antarctic. *Oceanol. Hydorbiol. Stud.* **2014**, *43*, 106–113. [CrossRef]
26. Barrera-Oro, E.R.; Moreira, E.; Seefeldt, M.A.; Francione, M.V.; Quartino, M.L. The importance of macroalgae and associated amphipods in the selective benthic feeding of sister rockcod species *Notothenia rossii* and *N. coriiceps* (Nototheniidae) in West Antarctica. *Polar Biol.* **2019**, *42*, 317–334. [CrossRef]
27. Gilles, C.L.; Stark, J.S.; Johnstone, G.J.; Smith, S.D.A. Carbon flow and trophic structure of an Antarctic coastal benthic community as determined by d13C and d15N. *Estuar. Coast. Shelf Sci.* **2012**, *97*, 44–57. [CrossRef]
28. Pasotti, F.; Saravia, L.A.; De Troch, M.; Tarantelli, M.S.; Sahade, R.; Vamreisel, A. Benthic trophic interactions in an Antarctic shallow water ecosystem affected by recent glacier retreat. *PLoS ONE* **2015**, *10*, e0141742. [CrossRef]
29. Servetto, N.; Rossi, S.; Fuentes, V.; Alurralde, G.; Lagger, C.; Sahade, R. Seasonal trophic ecology of the dominant Antarctic coral *Malacobelemnon daytoni* (Octacorallia, Pennatulacea, Kophobelemnidae). *Mar. Environ. Res.* **2017**, *130*, 264–274. [CrossRef]
30. Ha, S.-Y.; Ahn, I.-Y.; Moon, H.-W.; Choi, B.; Shin, K.-H. Tight trophic association between benthic diatom blooms and shallow-water megabenthic communities in a rapidly deglaciated Antarctic fjord. *Estuar. Coast. Shelf Sci.* **2019**, *218*, 258–267. [CrossRef]
31. Zenteno, I.; Cárdenas, I.; Valdivia, N.; Gómez, I.; Höfer, J.; Garrido, I.; Pardo, I.M. Unraveling the multiple bottom-up supplies of an Antarctic nearshore benthic community. *Prog. Oceanogr.* **2019**, *174*, 55–63. [CrossRef]
32. Richardson, M.G.; Whitaker, T.M. An Antarctic fast-ice food chain: Observations on the interaction of the amphipod *Pontogeneia antarctica* Chevreux with ice-associated microalgae. *Bull. Br. Antarct. Surv.* **1979**, *47*, 107–115.
33. Momo, F.R. Life Cycle and Spatial Distribution of *Gondogeneia antarctica* Chevreux (Crustacea, Amphipoda). Ph.D. Thesis, Universidad de Buenos Aires, Buenos Aires, Argentina, 1995. (In Spanish).
34. Amsler, C.; Iken, K.; McClintock, J.; Amsler, M.; Peters, K.; Hubbard, J.; Furrow, F.; Baker, B. Comprehensive evaluation of the palatability and chemical defenses of subtidal macroalgae from the Antarctic Peninsula. *Mar. Ecol. Prog. Ser.* **2005**, *294*, 141–159. [CrossRef]

35. Duarte, W.E.; Moreno, C.A. The specialized diet of *Harpagifer bispinis*: Its effect on the diversity of Antarctic intertidal amphipods. *Hydrobiologia* **1981**, *80*, 241–250. [CrossRef]
36. Creet, S.; Van Franeker, J.A.; Van Spanje, T.M.; Wolff, W.J. Diet of the Pintado petrel *Daption capense* at King George Island, Antarctica, 1990/91. *Mar. Ornithol.* **1994**, *22*, 221–229.
37. Fanta, E. Laboratory tests on feeding interactions and food preferences of some Antarctic fish from Admiralty Bay, King George Island, South Shetland Islands. *Pol. Polar Res.* **1999**, *20*, 335–346.
38. Barrera-Oro, E.R.; Piacentino, G.L.M. Feeding habits of juvenile *Trematomus newnesi* (Pisces, Nototheniidae) at Potter Cove, South Shetland Islands, Antarctica. *Polar Biol.* **2007**, *30*, 789–796. [CrossRef]
39. Barrera-Oro, E.R.; Winter, D.J. Age composition and feeding ecology of early juvenile *Notothenia rossii* (Pisces, Nototheniidae) at Potter Cove, South Shetland Islands, Antarctica. *Antarct. Sci.* **2008**, *20*, 339–341. [CrossRef]
40. Marina, T.I.; Salinas, V.; Cordone, G.; Campana, G.; Moreira, E.; Deregibus, D.; Torre, L.; Sahade, R.; Tatián, M.; Oro, E.B.; et al. The food web of Potter Cove (Antarctica): Complexity, structure and function. *Estuar Costal Shelf Sci.* **2018**, *200*, 141–151. [CrossRef]
41. Arrigo, K.R.; Worthen, D.; Schnell, A.; Lizotte, M.P. Primary production in Southern Ocean waters. *J. Geophys. Res.* **1998**, *103*, 15587–15600. [CrossRef]
42. Kim, D.-U.; Khim, J.S.; Ahn, I.-Y. Patterns, drivers and implications of ascidian distributions in a rapidly deglaciating fjord, King George Island, West Antarctic Peninsula. *Ecol. Indic.* **2021**, *125*, 107467. [CrossRef]
43. Chang, K.I.; Jun, H.K.; Park, G.T.; Eo, Y.S. Oceanographic conditions of Maxwell Bay, King George Island, Antarctica (austral summer 1989). *Korean J. Polar Res.* **1990**, *1*, 27–46.
44. Yoo, K.-C.; Lee, M.K.; Yoon, H.I.; Lee, Y.I.; Kang, C.Y. Hydrography of Marian Cove, King George Island, West Antarctica: Implications for ice-proximal sedimentation during summer. *Antarct. Sci.* **2015**, *27*, 185–196. [CrossRef]
45. Moon, H.W.; Wan Hussin, W.M.R.; Kim, H.C.; Ahn, I.Y. The impacts of climate change on Antarctic nearshore mega-epifaunal benthic assemblages in a glacial fjord on King George Island: Responses and implications. *Ecol. Indic.* **2015**, *57*, 280–292. [CrossRef]
46. Yoon, H.I.; Park, B.K.; Domack, E.W.; Kim, Y. Distribution and dispersal pattern of suspended particulate matter in Maxwell Bay and its tributary, Marian Cove, in the South Shetland Islands, West Antarctica. *Mar. Geol.* **1998**, *152*, 261–275. [CrossRef]
47. Yoo, K.; Yoon, H.; Oh, J.; Kim, Y.; Kang, C. Water column properties and dispersal pattern of suspended particulate matter (SPM) of Marian Cove during austral summer, King George Island, West Antarctica. *J. Korean Soc. Oceanogr.* **1999**, *4*, 266–274.
48. Yoo, K.; Kang, C.; Yoon, H.; Suk, D.; Oh, J. Seasonal water column properties and dispersal pattern of suspended particulate matter (SPM) in Marian Cove, King George Island, South Shetland Islands. *J. Korean Soc. Oceanogr.* **2002**, *38*, 573–593.
49. Ahn, I.Y.; Chung, K.H.; Choi, H.J. Influence of glacial runoff on baseline metal accumulation in the Antarctic limpet *Nacella concinna* from King George Island. *Mar. Pollut. Bull.* **2004**, *49*, 119–127. [CrossRef]
50. Chung, H.; Oh, Y.S.; Lee, I.K.; Kim, D.Y. Macroalgal vegetation of Maxwell Bay in King George Island, Antarctica. *Korean J. Phycol.* **1994**, *9*, 47–58.
51. Kim, J.H.; Chung, H.; Oh, Y.S.; Lee, I.K. Macroalgal flora of Maxwell Bay, King George Island, Antarctica: I. Chlorophyta, Chrysophyta and Phaeophyta. *Ocean. Polar Res.* **2001**, *23*, 209–221.
52. Kim, J.H.; Chung, H.; Oh, Y.S.; Lee, I.K. Macroalgal flora of Maxwell Bay, King George Island, Antarctica: II. Rhodophyta. *Ocean. Polar Res.* **2001**, *23*, 347–360.
53. Ko, Y.W.; Choi, H.-G.; Lee, D.S.; Kim, J.H. 30 years revisit survey for long-term changes in the Antarctic subtidal algal assemblage. *Sci. Rep.* **2020**, *10*, 8481. [CrossRef]
54. Iken, K.; Barrera-Oro, E.R.; Quartino, M.L.; Casaux, R.J.; Brey, T. Grazing by the Antarctic fish *Notothenia coriiceps*: Evidence for selective feeding on macroalgae. *Antarct. Sci.* **1997**, *9*, 386–391. [CrossRef]
55. Iken, K.; Quartino, M.; Wiencke, C. Histological identification of macroalgae from stomach contents of the Antarctic fish *Notothenia coriiceps* using semi-thin sections. *Mar. Ecol.* **1999**, *20*, 11–17. [CrossRef]
56. Bucolo, P.; Amsler, C.D.; McClintock, J.B.; Baker, B.J. Palatability of the Antarctic rhodophyte *Palmaria decipiens* (Reinsch) RW Ricker and its endo/epiphyte *Elachista antarctica* Skottsberg to sympatric amphipods. *J. Exp. Mar. Biol. Ecol.* **2011**, *396*, 202–206. [CrossRef]
57. Doyle, S.R.; Momo, F.R.; Brêthes, J.C.; Ferreyra, G.A. Metabolic rate and food availability of the Antarctic amphipod *Gondogeneia antarctica* (Chevreux 1906): Seasonal variation in allometric scaling and temperature dependence. *Polar Biol.* **2012**, *35*, 413–424. [CrossRef]
58. Ahn, I.Y.; Moon, H.W.; Jeon, M.; Kang, S.H. First record of massive blooming of benthic diatoms in their association with megabenthic filter feeders on the shallow seafloor of an Antarctic fjord: Does glacier melting fuel the bloom? *Ocean Sci. J.* **2016**, *51*, 273–279. [CrossRef]
59. Dunton, K.H. δ^{13}C and δ^{15}N measurements of Antarctic Peninsula fauna: Trophic relationships and assimilation of benthic seaweeds. *Am. Zool.* **2001**, *41*, 99–112. [CrossRef]
60. Ahn, I.Y.; Chung, H.; Kang, J.S.; Kang, S.H. Diatom composition and biomass variability in nearshore waters of Maxwell Bay, Antarctica during the 1992/1993 austral summer. *Polar Biol.* **1997**, *17*, 123–130. [CrossRef]
61. Choy, E.J.; Park, H.; Kim, J.H.; Ahn, I.Y.; Kang, C.K. Isotopic shift for defining habitat exploitation by the Antarctic limpet *Nacella concinna* from rocky coastal habitats (Marian Cove, King George Island). *Estuar. Coast. Shelf Sci.* **2011**, *92*, 339–346. [CrossRef]

62. Jacob, U.; Brey, T.; Fetzer, I.; Kaehler, S.; Mintenbeck, K.; Dunton, K.; Beyer, K.; Struck, U.; Pakhomov, E.; Arntz, W.E. Towards the trophic structure of the Bouvet Island marine ecosystem. *Polar Biol.* **2005**, *29*, 106–113. [CrossRef]
63. Bunn, S.E.; Loneragan, N.R.; Kempster, M.A. Effects of Acid Washing on Stable Isotope Ratios of C and N in Penaeid Shrimp and Seagrass: Implications for Food-Web Studies Using Multiple Stable Isotopes. *Limnol. Oceanogr.* **1995**, *40*, 622–625. [CrossRef]
64. Logan, J.M.; Lutcavage, M.E. A comparison of carbon and nitrogen stable isotope ratios of fish tissues following lipid extractions with non-polar and traditional chloroform/methanol solvent systems. *Rapid Commun. Mass Spectrom.* **2008**, *22*, 1081–1086. [CrossRef]
65. Soler-Membives, A.; Rossi, S.; Munilla, T. Feeding ecology of *Ammothella longipes* (Arthropoda: Pycnogonida) in the Mediterranean Sea: A fatty acid biomarker approach. *Estuar. Coast. Shelf Sci.* **2011**, *92*, 588–597. [CrossRef]
66. Gori, A.; Rossi, S.; Linares, C.; Berganzo, E.; Orejas, C.; Dale, M.R.T.; Gili, J.M. Size and spatial structure in deep versus shallow populations of the Mediterranean gorgonian *Eunicella singularis* (Cap de Creus, northweste.rn Mediterranean Sea). *Mar. Biol.* **2011**, *158*, 1721–1732. [CrossRef]
67. Viladrich, N.; Bramanti, L.; Tsounis, G.; Chocarro, B.; Martínez-Quintana, A.; Ambroso, S.; Madurell, T.; Rossi, S. Energetic resource allocation in reproduction for two temperate octocorals with contrasting reproductive strategies: Surface versus internal brooder. *Coral Reefs* **2016**, *35*, 1033–1045. [CrossRef]
68. Clarke, K.R.; Gorley, R.N. *PRIMER 6 User Manual/Tutorial*; PRIMER-E Ltd.: Plymouth, UK, 2006.
69. KOPRI. *Overwintering Report of the Antarctic King Sejong Station 28th Overwintering Party (December 2014–December 2015)*; Korea Polar Research Institute: Incheon, Korea, 2016; p. 881.
70. DeNiro, M.J.; Epstein, S. Influence of diet on the distribution of carbon isotopes in animals. *Geochim. Cosmochim Acta* **1978**, *42*, 495–506. [CrossRef]
71. Aumack, C.F.; Amsler, C.D.; Mcclintock, J.B.; Baker, B.J. Chemically mediated resistance to mesoherbivory in finely branched macroalgae along the western Antarctic Peninsula. *Eur. J. Phycol.* **2010**, *45*, 19–26. [CrossRef]
72. Quartino, M.L.; Boraso de Zaixso, A.L. Summer macroalgal biomass in Potter Cove, South Shetland Islands, Antarctica: Its production and flux to the ecosystem. *Polar Biol.* **2008**, *31*, 281–294. [CrossRef]
73. Wiencke, C. Seasonality of Red and Green Macroalgae from Antarctica—A Long-Term Culture Study under Fluctuating Antarctic Daylengths. *Polar Biol.* **1990**, *10*, 601–607. [CrossRef]
74. Weykam, G.; Wiencke, C. Seasonal photosynthetic performance of the endemic antarctic red alga *Palmaria decipiens* (Reinsch) Ricker. *Polar Biol.* **1996**, *16*, 357–361. [CrossRef]
75. Weykam, G.; Thomas, D.N.; Wiencke, C. Growth and photosynthesis of the Antarctic red algae *Palmaria decipiens* (Palmariales) and *Iridaea cordata* (Gigartinales) during and following extended periods of darkness. *Phycologia* **1997**, *36*, 395–405. [CrossRef]
76. Dunton, K.H.; Schell, D.M. Dependence of consumers on macroalgal (*Laminaria solidungula*) carbon in an arctic kelp community: δ^{13}C evidence. *Mar. Biol.* **1987**, *93*, 615–625. [CrossRef]
77. Mintenbeck, K.; Jacob, U.; Knust, R.; Arntz, W.E.; Brey, T. Depth-dependence in stable isotope ratio δ^{15}N of benthic POM consumers: The role of particle dynamics and organism trophic guild. *Deep Sea Res.* **2007**, *54*, 1015–1023. [CrossRef]
78. Roy, V.; Iken, K.; Gosselin, M.; Tremblay, J.É.; Bélanger, S.; Archambault, P. Benthic faunal assimilation pathways and depth-related changes in food-web structure across the Canadian Arctic. *Deep Sea Res.* **2015**, *102*, 55–71. [CrossRef]
79. Braeckman, U.; Pasotti, F.; Vázquez, S.; Zacher, K.; Hoffmann, R.; Elvert, M.; Marchant, H.; Buckner, C.; Quartino, M.L.; Mác Cormack, W.; et al. Degradation of macroalgal detritus in shallow coastal Antarctic sediments. *Limnol. Oceanogr.* **2019**, *64*, 1423–1441. [CrossRef] [PubMed]
80. Mincks, S.; Smith, C.R.; DeMaster, D.J. Persistence of labile organic matter and microbial biomass in Antarctic shelf sediments: Evidence of a sediment 'food bank'. *Mar. Ecol. Prog. Ser.* **2005**, *300*, 3–19. [CrossRef]
81. Fischer, G.; Wiencke, C. Stable carbon isotope composition, depth distribution and fate of macroalgae from the Antarctic Peninsula region. *Polar Biol.* **1992**, *12*, 341–348. [CrossRef]
82. Corbisier, T.N.; Petti, M.A.V.; Skowronski, R.S.P.; Brito, T.A.S. Trophic relationships in the nearshore zone of Martel Inlet (King George Island, Antarctica): d13C stable-isotope analysis. *Polar Biol.* **2004**, *27*, 75–82. [CrossRef]
83. Kaehler, S.; Pakhomov, E.A.; Kalin, R.M.; Davis, S. Trophic importance of kelp-derived suspended particulate matter in a through-flow sub-Antarctic system. *Mar. Ecol. Prog. Ser.* **2006**, *316*, 17–22. [CrossRef]
84. Lastra, M.; Rodil, I.F.; Sánchez-Mata, A.; García-Gallego, M.; Mora, J. Fate and processing of macroalgal wrack subsidies in beaches of Deception Island, Antarctic Peninsula. *J. Sea Res.* **2014**, *88*, 1–10. [CrossRef]
85. Amsler, C.D.; McClintock, J.B.; Baker, B.J. Palatability of living and dead detached Antarctic macroalgae to consumers. *Antarct. Sci.* **2012**, *24*, 589–590. [CrossRef]
86. Ziegler, A.F.; Cape, M.; Lundesgaard, Ø.; Smith, C.R. Intense deposition and rapid processing of seafloor phytodetritus in a glaciomarine fjord, Andvord Bay (Antarctica). *Prog. Oceanogr.* **2020**, *187*, 102413. [CrossRef]
87. Dunstan, G.A.; Volkman, J.K.; Barrett, S.M.; Leroi, J.-M.; Jeffrey, S.W. Essential polyunsaturated fatty acids from 14 species of diatom (bacillariophyceae). *Phytochemistry* **1994**, *35*, 155–161. [CrossRef]
88. Kelly, J.; Scheibling, R. Fatty acids as dietary tracers in benthic food webs. *Mar. Ecol. Prog. Ser.* **2012**, *446*, 1–22. [CrossRef]
89. Graeve, M.; Dauby, P.; Scailteur, Y. Combined lipid, fatty acid and digestive tract content analyses: A penetrating approach to estimate feeding modes of Antarctic amphipods. *Polar Biol.* **2001**, *24*, 853–862.

90. Lagger, C.; Nime, M.; Torre, L.; Servetto, N.; Tatián, M.; Sahade, R. Climate change, glacier retreat and a new ice-free island offer new insights on Antarctic benthic responses. *Ecography* **2018**, *41*, 579–591. [CrossRef]
91. Quartino, M.L.; Deregibus, D.; Campana, G.L.; Latorre, G.E.J.; Momo, F.R. Evidence of macroalgal colonization on newly ice-free areas following glacial retreat in Potter Cove (South Shetland Islands), Antarctica. *PLoS ONE* **2013**, *8*, e58223. [CrossRef]
92. Schram, J.B.; Schoenrock, K.M.; McClintock, J.B.; Amsler, C.D.; Angus, R.A. Ocean warming and acidification alter Antarctic macroalgal biochemical composition but not amphipod grazer feeding preferences. *Mar. Ecol. Prog. Ser.* **2017**, *581*, 45–56. [CrossRef]

Journal of
Marine Science and Engineering

MDPI

Article

Holoplanktonic and Meroplanktonic Larvae in the Surface Waters of the Onnuri Vent Field in the Central Indian Ridge

Minju Kim [1,2], Jung-Hoon Kang [1,2,*] and Dongsung Kim [3]

1 Risk Assessment Research Center, Korea Institute of Ocean Science and Technology, Geoje 53201, Korea; minjukim@kiost.ac.kr
2 Department of Ocean Science, University of Science and Technology, Daejeon 34113, Korea
3 Marine Ecosystem Research Center, Korea Institute of Ocean Science and Technology, Busan 49111, Korea; dskim@kiost.ac.kr
* Correspondence: jhkang@kiost.ac.kr

Abstract: The dispersal of organisms in an isolated environment of a hydrothermal vent remains unclear. Here, we provide direct evidence that meroplanktonic larvae may migrate thousands of meters above the ocean floor. The morphological quantitative measurements of mesozooplankton were conducted in the Onnuri Vent Field (OVF), the Solitaire Fields (SF), and the reference site (ref-site). Only one species of bivalve larva that appeared at the OVF and the ref-site (0–200 m) was similar to *Bathymodiolus* spp. Sixteen species of gastropod larvae were distinguished, among which, species 1–4, 6, and 13 had holoplanktonic features (*Atlanta, Oxygrus,* and *Limacina*), whereas species 5, 7–12, and 14–16 had meroplanktonic features. Species 5, 11, and 12 appeared only at the OVF, 9 and 10 appeared only in the SF, 14–16 appeared only at the ref-site, and species 7 and 8 appeared in all surveyed stations. The species 5, 8, 12, 14, and 15 have morphological features similar to *Vetulonia* spp., and species 7 was similar to Lepetodrilidae; species 9–11 and 16 were similar to *Phymorphynchus protoconchs.* The morphologically distinguished mollusk larvae in the upper layers of the water column (0–200 m) indicate that larvae associated with deep-sea hydrothermal vents may disperse approximately 2000 m above the vents.

Keywords: hydrothermal vent; mesozooplankton; meroplankton; mollusk larvae; bivalve larvae; gastropod larvae; dispersal; Central Indian Ridge; Onnuri Vent Field

1. Introduction

Benthic animals produce planktotrophic larvae, which spend weeks to months in the pelagic environment, feeding and developing through their free-swimming stage [1–4]. Larval dispersal is important for geographic distribution, population dynamics, and evolutionary processes of vent-endemic organisms [1,5,6]. Planktonic larval duration (PLD) is the amount of time larvae take to develop to the settlement stage [5]. PLD and dispersal strategies may differ depending on the species, climate zone, and habitat types [7]. In the near-shore ecosystems, the species living on sandy bottoms have long-distance dispersal mechanisms, whereas the rocky-shore species that depend on a narrow strip of habitat have short-distance dispersal mechanisms to reduce offshore loss of larvae [8]. Shanks (2009) reported that shallow-water and coastal species are more concentrated in complex hydrodynamic and fast-speed currents than in the deep-sea. The intertidal and shallow subtidal species are transported from inshore waters into large-scale coastal currents and then eventually return to the inshore waters of another location [9]. Deep-sea species also have complicated PLD, where larval transport distance depends on the development, behavior, and physiology of the larvae [5]. For example, veliger larvae such as limpets (e.g., *Leptodrilus* spp.) tend to remain near the bottom and in slower currents [10]; tubeworms (e.g., *Riftia pachyptila*) are buoyant and more likely to be transported above the bottom (e.g., ridge-controlled currents) [11]; and swimming larvae such as mussels (e.g., *Bathymodiolus childressi*) migrate up into the oceanic currents near the surface [12].

The hydrothermal vent fields have distinct and isolated communities with chemosynthetic systems [13]. The majority of hydrothermal vent sites, particularly in the Pacific and Atlantic Oceans, have been studied [14,15]. A hydrothermal vent has either a discrete or a diffuse venting source, such as black and white smokers [16–18]. Another type of vent field is the diffused flow from porous surfaces or fissures and cracks in basalt lavas, diluted with cold seawater either from below or within the sea floor [19]. The benthic animals inhabiting the hydrothermal vent ecosystems are likely to have colonization abilities, including high rates of dispersal, growth, and reproduction [20]. The major vent fauna in the extensively studied Galapagos vents, including tubeworms (Alvinellidae), clams (e.g., *Calyptogena magnifica*), and mollusks (e.g., *B. thermophilus*), have symbiotic bacteria that obtain energy from hydrogen sulfide [21].

Most species in hydrothermal vents are benthic as adults, but they spend a portion of their lives as tiny larvae (meroplankton) in the overlying water column [5]. Previous studies on zooplankton, including meroplankton, in the water column above the hydrothermal vents were conducted in the East Pacific Rise (EPR), Guaymas Basin, Endeavour Ridge, Juan de Fuca Ridge, and Mid-Atlantic Ridge [22–26]. The larval stage (meroplankton) of macrobenthic animals living near the hydrothermal vent is an important phase of their life cycle [20,27]. The larval dispersal, especially migration behaviors of benthic organisms in hydrothermal vents has primarily been studied at EPR and Juan de Fuca Ridge [28–31]. Hydrodynamics and chemical and mass flux studies in the EPR demonstrated the transport mechanisms of vent fluids and larvae, implying eddy-driven impacts in transporting hydrothermal vent-derived heat, chemicals, and biota in a relatively low-energy environment [32].

Meroplanktonic larvae of mollusks in the cold methane seeps, including mytilid bivalve *B. childressi* and limpet gastropod *Bathynerita naticoidea*, have been collected from euphotic zones (0–100 m depth), implying that they migrated to surface waters above the ocean floor [33]. Yahagi et al. (2017) provided evidence of early life history traits and population genetics for the surface dispersal of vent species of the gastropod *Shinkailepas myojinensis* that inhabited sulfide chimneys and volcanic rocks covered by bacterial mats in diffuse venting areas. The hatched larvae swam upwards depending on temperature, and this migration was probably because of the high food source (high phytoplankton biomass) that helps in their growth and development [6].

In this study, we attempted to present the vertical distribution of the holoplanktonic and meroplanktonic larvae in the hydrothermal vents of the Central Indian Ridge. To determine the difference in distribution patterns of larvae from the deep to the sea surface in the diffuse type (Onnuri Vent Field, OVF) and chimney type vent (Solitaire Field, SF), we collected larvae using a multiple opening and closing net environmental sensing system (MOCNESS).

2. Materials and Methods

2.1. Study Area and Physicochemical Analysis

The OVF (newly discovered; 11°23.86′ S, 66°24.42′ E) and SF (19°33.4′ S, 65°50.9′ E) were the two vent fields selected in the Central Indian Ridge (CIR) to be investigated during a cruise of the R/V ISABU in June 2018 (Figure 1A,B). The CIR is known to be slow to intermediate spreading, with spreading rates varying from <12 mm/year to 50–60 mm/year [34]. The OVF is located at the summit of the Ocean Core Complex-3-2 and vents low-temperature fluids at depths of 1990–2170 m [35]. The SF is located on the Roger plateau on the western ridge of CIR segment 15 at a depth of 2634 m [15,34]. SF is characterized by chimneys that are less than 5 m in height and with fluid emissions of high pH 4.8 and approximately 10% lower chlorine levels than the ambient seawater [15]. The reference site (ref-site, MOCNESS tow: 09°22.4′–9°27.7′ S, 67°10.8′–67°15.8′ E) was selected outside of the immediate influence of the hydrothermal vent fields. The ideal location was north of OVF as the direction of transport of meridional Ekman flows from north to south during the boreal summer monsoon in June [36].

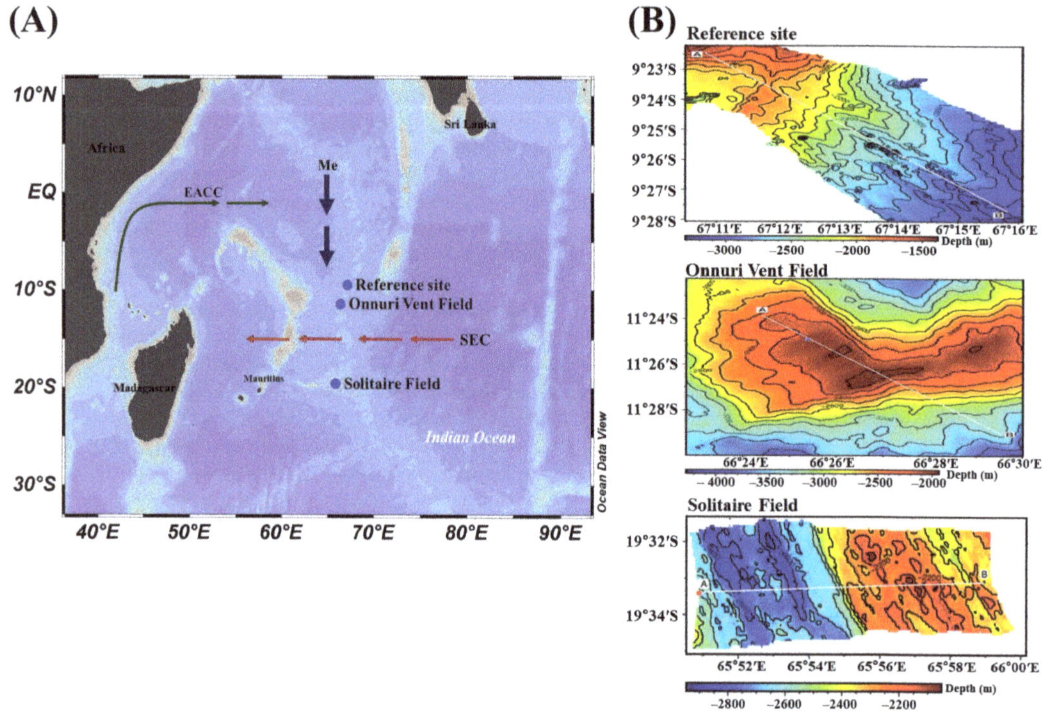

Figure 1. (**A**) Map showing the selected survey stations at the Central Indian Ridge in June 2018. Surface currents indicated are the South Equatorial Current (SEC), East Africa Coastal Current (EACC), and the direction of meridional Ekman transports (Me). The major currents are cited from Hood et al. (2017). (**B**) The figure on the right shows bathymetry from the start to the end of the multiple opening and closing net environmental sensing system (MOCNESS) sampling in both study stations.

The physical properties were obtained from the conductivity (#217), temperature (#6120), and depth (#217) sensors (Sea-bird Electronics, Bellevue, WA, USA) attached to MOCNESS (Biological environmental sampling systems Inc., BESS, North Falmouth, MA, USA). PVC Niskin bottles mounted on an instrumented Rosette sampler were used to collect discrete seawater samples for chlorophyll-*a* (chl-*a*) concentration analysis. Due to limited sampling time, water samples for chl-*a* concentration analysis were not collected at the ref-site. Seawater samples (2 L) for analysis of chl-*a* concentration were collected at 5 depths, including surface and subsurface chlorophyll maximum (SCM) layers, from 0–0.5 m (surface) to 200 m at OVF and SF. Seawater samples were filtered using 47 mm GF/F glass-fiber filters (Whatman, GE Healthcare, Chicago, IL, USA) under low vacuum pressure (<125 mmHg). The filters were kept frozen in a deep freezer (−80 °C) before being extracted in 90% acetone at 4 °C in the dark for 24 h. Chl-*a* concentration was determined using Turner Designs 10-AU, which was calibrated using a chl-*a* standard solution (Turner Designs, San Jose, CA, USA) based on the method by Parsons et al. (1984) [37].

2.2. Mesozooplankton Including Larvae Sampling

The mesozooplankton samples were collected using a 1 m^2 MOCNESS equipped with nine nets (mesh size 200 μm). To minimize sample contamination, net traps and bar stops were employed to restrict net bar movement when the nets were opened or closed. The system carried sensors to measure pressure, temperature, and conductivity (Sea-Bird Electronics, Bellevue, WA, USA). The MOCNESS was attached to the deep-tow winch on the R/V ISABU, and the speeds of the winch wire during pay-out and haul-in were 20–30 m min^{-1}. The speed of the vessel was between 1.5 and 2.0 knots during towing. The net angle was maintained between 35° and 55° (optimum angle: 45°; Biological Environmental Sampling Systems Inc., BESS, North Falmouth, MA, USA) by increasing or decreasing the speed of either the ship or the winch wire. Full depth sample collection was obtained by oblique tows from 100 m above the bottom of the sea. The depth strata sampled were divided into 6–7 layers (Table 1). The volume of filtered water was calculated using a flowmeter mounted on the frame of MOCNESS (Biological Environmental Sampling Systems Inc., BESS, North Falmouth, MA, USA). Mesozooplankton samples in the cod-end bucket were transferred to 1 L sampling bottles. The samples were divided into two groups, one sample group was immediately fixed to a final concentration of 5% with borax-neutralized formalin for microscopic examination, and the other was fixed with ethyl-alcohol (99.9%) for further molecular analysis. Mesozooplankton were identified into the lowest taxonomic groups, generally species or genus, and identified and enumerated under a stereomicroscope at 10×–80× magnification (Discovery V8, SteREO, Zeiss, Germany) in accordance with the studies of Conway et al. (2003) [38], and Chihara and Murano (1997) [39]. The larvae of benthic invertebrates at the hydrothermal vents were identified using the photographic identification guide to larvae at hydrothermal vents by Mills et al. (2009) [27]. The gastropod larvae were distinguished and divided into holoplanktonic and meroplanktonic larvae by their morphological characteristics, which were analyzed using the photographic guide. The photographs of mollusks were taken by the stereomicroscope equipped with the camera at 25×, 32×, 50×, 63×, and 80× magnifications (AxioCam ICc 3, Zeiss, Germany). The number of individuals was converted to individuals per cubic meter (ind. 100 m^{-3}). The species richness (Margalef species richness index) and diversity index (Shannon-Wiener diversity index) were calculated by using PRIMER V6.1.10 statistical package (PRIMER-E, Plymouth Marine Laboratory, Plymouth, United Kingdom).

Table 1. The sampling strata (depth, m), and average for temperature (temp.), and salinity (sal.), integrated chlorophyll-*a* (chl-*a*), total abundance of zooplankton (zoopl.), and top three dominant zooplankton (zoopl.) are listed for each station during the study period. N.S.: not sampled.

Station	Sampling Strata (m)	Temp. (°C)	Sal.	Chl-*a* (mg m^{-3})	Total Zoopl. (ind. 100 m^{-3})	Top Three Dominant Zoopl. (Relative Abundance)
	(1) 0–200	23.7	35.2	3.1	19,522	*Oithona* copepodites (28%), Radiolarians (8%), *Pleuromamma* copepodites (6%)
	(2) 200–500	15.7	35.4	N.S.	4542	*Pleuromamma* copepodites (12%), *Oithona* copepodites (11%), ostracod sp.1 (11%)
Solitaire Field	(3) 500–1000	7.9	34.6	N.S.	1291	*Oncaea* copepodites (26%), *Pleuromamma* copepodites (23%), *Oithona* copepodites (10%)
	(4) 1000–1500	4.4	34.6	N.S.	501	Radiolarians (18%), *Oithona* copepodites (18%), *Oncaea* copepodites (11%)
	(5) 1500–2000	2.7	34.7	N.S.	171	Radiolarians (27%), *Oncaea* copepodites (18%), *Oithona fallax* (18%)
	(6) 2000–2500	2.0	34.7	N.S.	60	Radiolarians (16%), *Oncaea* copepodites (16%), *Microsetella* spp. (16%)
	(1) 0–200	18.3	35.0	13.5	71,308	*Paracalanus* copepodites (16%), *Pyrocystis noctiluca* (11%), foraminiferans (7%)
	(2) 200–500	10.7	34.8	N.S.	4474	Radiolarians (19%), *Oncaea* copepodites (13%), *Scolecithrix* copepodites (13%)
Onnuri Vent Field	(3) 500–1000	6.9	34.7	N.S.	2850	*Oncaea* copepodites (32%), Harpacticoid copepodites (15%), *Oithona* copepodites (10%)
	(4) 1000–1300	4.9	34.7	N.S.	922	*Oncaea* copepodites (25%), Radiolarians (16%), *Scolecithrix* copepodites (13%)
	(5) 1300–1600	3.8	34.7	N.S.	1138	*Oncaea* copepodites (43%), Radiolarians (28%), Harpacticoid copepodites (8%)
	(6) 1600–1900	3.0	34.7	N.S.	19	*Oncaea* copepodites (42%), Harpacticoid copepodites (13%), foraminiferans (8%)
	(1) 0–200	20.9	34.8	N.S.	77,037	*Pyrocystis noctiluca* (14%), *Globigerina* spp. (7%), *Oncaea* copepodites (7%)
Reference site	(2) 200–400	11.8	34.9	N.S.	8430	*Oithona* copepodites (18%), *Oncaea* copepodites (14%), *Pleuromamma* copepodites (6%)
	(3) 400–1000	7.4	34.7	N.S.	11,460	*Oncaea* copepodites (25%), *Oithona* copepodites (12%), *Scolecithrix* copepodites (10%)
	(4) 1000–1300	5.1	34.7	N.S.	483	*Oncaea* copepodites (46%), *Oithona* copepodites (12%), *Scolecithrix* copepodites (6%)

3. Results

3.1. Physicochemical Properties and Chlorophyll-a Concentrations

The vertical distributions of temperature and salinity in the water columns were acquired at all survey stations. The temperature observation showed that the depth of the surface mixed layer (SML) increased with an increase in latitude (°S) (avg. 79.2 m) (Figure 2). The deepest SML depth was observed in the SF (100 m; with an average temperature of 25.1 °C), and the shallowest SML depth was seen in the ref-site (63.0 m; with an average temperature of 27.2 °C). The SML depth for OVF was 74.5 m, with an average temperature of 27.0 °C. Warmer surface waters (avg. 27.2 °C) were observed in the ref-site and OVF, while relatively cooler surface water (avg. 25.1 °C) was observed in the SF. The temperature stratification in the SF and OVF was stronger from the surface to 350 m, and weakly stratified with decreasing depth. At SF and OVF, the bottom temperature was 2.7 °C (1900 m) and 4.5 °C (1300 m), respectively. The thermocline (temperature decreases rapidly from the SML) existed between 100–900 m, below which the temperature remained constant (avg. 3.2 °C) and reached 1.9 °C near the bottom (2500 m). The surface water in SF was relatively more saline (avg. 35.0) than the surface waters in OVF and ref-site (avg. 34.5). The salinity remained constant from surface to SML with increasing depth in all surveys, except between a layer of SML and 100 m. The integrated chl-a concentration within the SML was also higher in OVF (4.0 mg m^{-2}) than in SF (1.8 mg m^{-2}) (Figure 2).

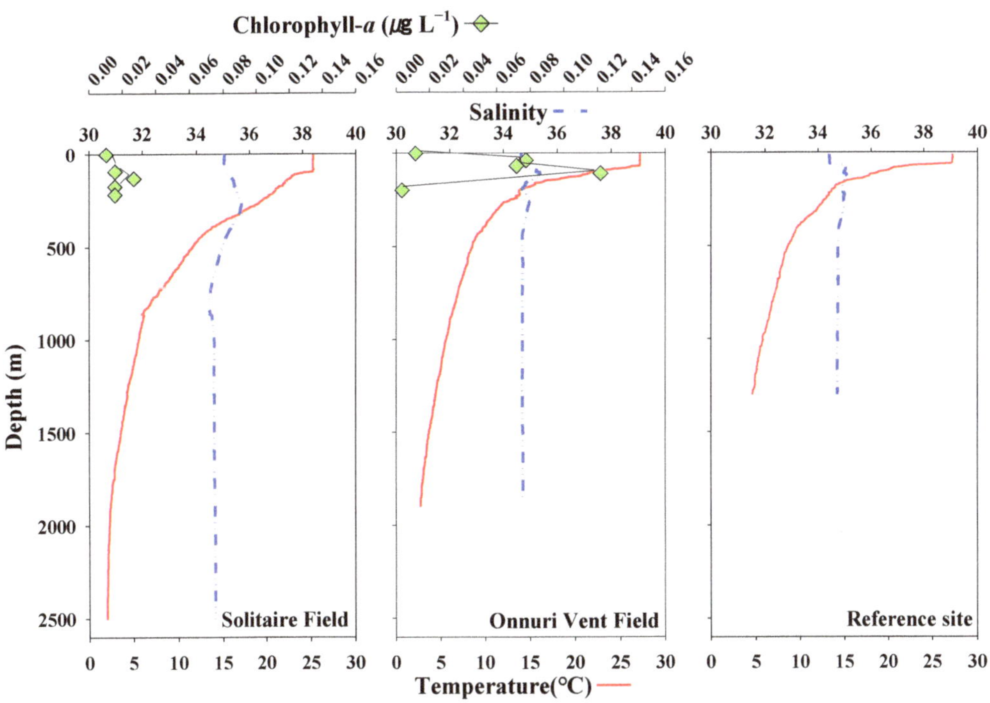

Figure 2. The vertical profiles of temperature (°C), salinity, and chlorophyll-a (µg L^{-1}) of the surveyed stations at the Central Indian Ridge during a cruise of the R/V ISABU in June 2018.

3.2. Abundance and Taxonomic Groups of Mesozooplankton

The total mesozooplankton abundance decreased with the depth, ranging from 60 to 19,522 ind. 100 m^{-3} (avg. 4348 ind. 100 m^{-3}) in SF; 19 to 71,308 ind. 100 m^{-3} (avg. 13,452 ind. 100 m^{-3}) in OVF; and 483 to 77,037 ind. 100 m^{-3} (avg. 24,352 ind. 100 m^{-3}) in the ref-site (Figure 3). It is evident that the majority of the mesozooplankton abundance is accumulated at 0–200 m (euphotic zone) in both water columns above hydrothermal vent fields, which may be due to higher chl-*a* concentrations in the SCM (avg. 112.5 m). The abundance at 0–200 m was approximately 11 times higher than in other layers (200–1300 m) at the ref-site.

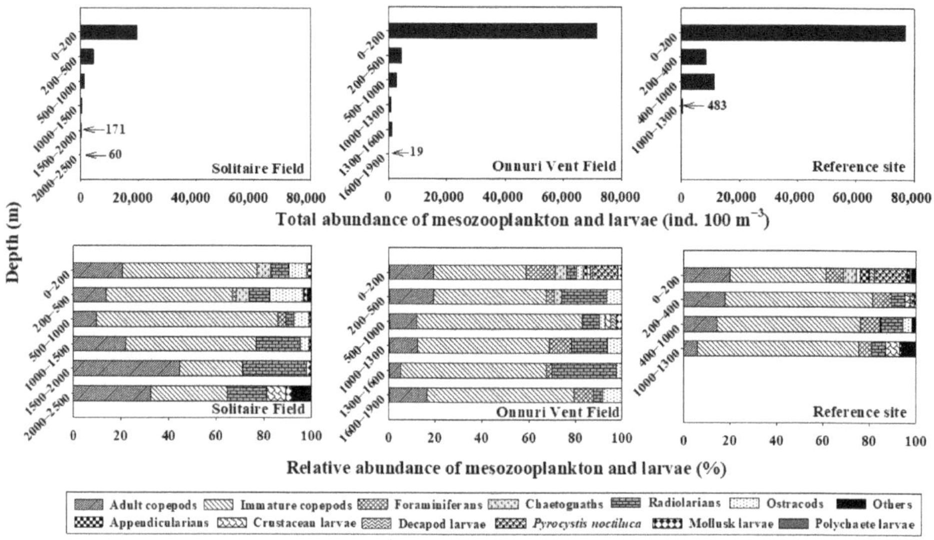

Figure 3. The vertical distribution of total and relative abundance of the mesozooplankton community in the surveyed stations during a cruise of the R/V ISABU in June 2018 at the Central Indian Ridge.

Most of the mesozooplankton community was composed of copepods (immature and adults) for all water columns of the surveys (Figure 3, Table 1). At the SF, copepods (both immature and adults) accounted for 65–86% (avg. 74%) of the total mesozooplankton community, followed by radiolarians that accounted for 3–27% (avg. 13%) occurring throughout the water column (0–2500 m), and ostracods and chaetognaths primarily occurring in the upper layers (0–500 m). Taxonomic groups of crustacean larvae and others included amphipod larvae and unidentified eggs, which appeared in the bottom layers (2000–2500 m). At the OVF, copepods (both immature and adults) accounted for 59–83% (avg. 71%) of the total mesozooplankton community, followed by radiolarians (avg. 13%), foraminiferans (avg. 6%), and ostracods (avg. 5%), occurring in all water columns. At the ref-site, copepods (immature and adult) accounted for 61–81% (avg. 74%) of the total mesozooplankton community, followed by foraminiferans (7%), radiolarians (5%), and *Pyrocystis noctiluca* (4%), occurring in all water columns. *P. noctiluca* accounted for 11% and 14% of the total mesozooplankton community at OVF and the ref-site, respectively, occurring between 0–200 m, probably due to relatively warmer surface waters (avg. 27.2 °C) (Figure 3, Table 1).

The morphologically identified numbers in the integrated water column were 28, 45, and 52 species in 6, 6, and 8 phyla at SF, OVF and the ref-site, respectively (Figure 4). The phyla Retaria (Foraminifera), Protozoa (Dinoflagellata), Crustacea (Ostracoda, Calanoida, Cyclopoida, Poecilostomatoida, Harpacticoida, Amphipoda, Euphausiacea, Decapoda), Chaetognatha (Sagittoidae), Chordata (Thaliacea, Doliolida, Appendicularia, Ichtyoplankton), and Mollusca (Gastropoda, Bivalvia) commonly appeared at both hydrothermal vents and the ref-site, while phyla Cnidaria (Hydrozoa) and Annelida (Polychaeta) only appeared at the ref-site (Figures 3 and 4). Both the richness and diversity of the species were highest at the ref-site (richness: 4.8, diversity: 3.1) and lowest at SF (richness: 2.9, diversity: 2.6).

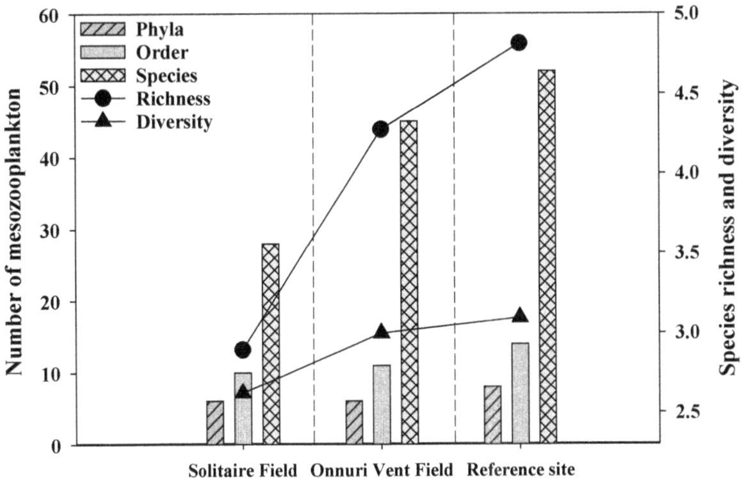

Figure 4. The number of mesozooplankton phyla, order, and species, and indices of species richness and diversity in the surveyed stations during a cruise of the R/V ISABU in June 2018 at the Central Indian Ridge.

Gastropod larvae were the dominant larvae at both SF (avg. 88%) and OVF (avg. 56%) (Figure 3), followed by amphipod larvae (avg. 12%) at SF and decapod larvae (avg. 17%) at OVF (Figure 3). Copepod nauplii (avg. 49%) was the most dominant larval group at the ref-site, which was followed by gastropod larvae (avg. 27%). We examined the taxonomic groups of mollusk larvae in further detail because deep-sea hydrothermal vent communities may disperse through planktotrophic modes (Figure 5) [1]. Mollusk larvae (bivalves and gastropods) can be holoplankton (spend entire life as plankton) or meroplankton (spend part of their life as plankton). We identified 18 species of mollusk larvae, including 16 unknown species of gastropod larvae, one species of bivalve larvae, and one species of pteropod larvae (*Cresis* spp.) (Figure 5).

The bivalve larvae (length: 330 μm) were found at both the OVF and the ref-site between depths of 0–200 m (Figure 5). The bivalve larvae had morphological features similar to those of the planktotrophic larvae of *Bathymodiolus* spp. that were reported to be found at hydrothermal vents in the South Pacific Ocean [27,40]. Gastropod larval species of 1–4, 6, and 13 have morphological features of holoplanktonic larvae such as pelagic species of the genus *Atlanta, Oxygrus,* and *Limacina* [38]. Gastropod larval species of 5, 7–12, and 14–16 have morphological features of meroplankton but only appear at depths between the surface and 200 m (Figure 5). Gastropod larval species 5, 11, and 12 only appeared at the OVF; 9 and 10 only appeared at the SF; 14–16 appeared only at the ref-site; 7–8 appeared in all surveyed stations. Gastropod larval species 5, 8, 12, 14, and 15 have similar morphological features to macrobenthic gastropod larvae of *Vetulonia* spp. Gastropod larvae species 7 have morphological features similar to gastropod Lepetodrilidae (*Lepetodrilus* spp.). Gastropod larval species 9–11 and 16 have morphological features similar to those of *Phymorphynchus protoconchs* [27].

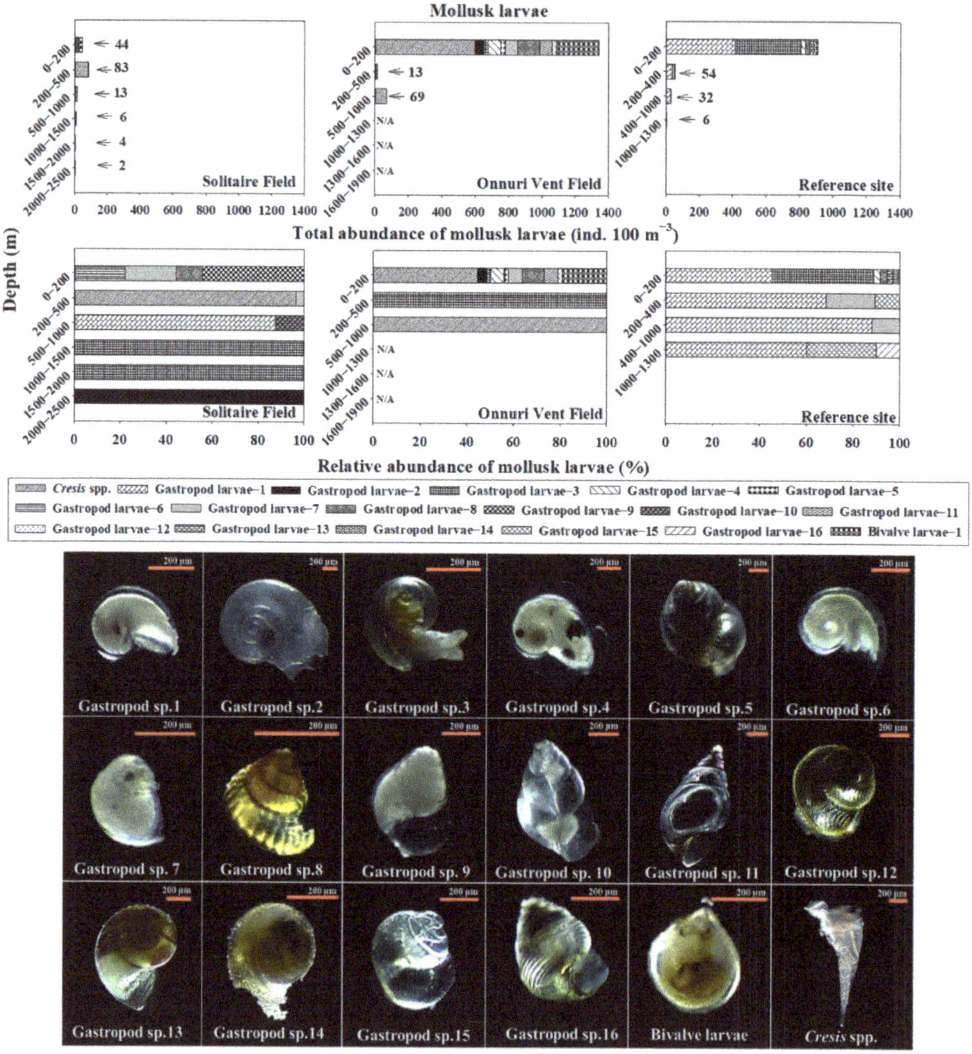

Figure 5. The vertical distribution of the total and relative abundance of mollusk larvae in the surveyed stations during a cruise of the R/V ISABU in June 2018 at the Central Indian Ridge. The 18 species of mollusk larvae, include 16 unknown species of gastropod larvae, one species of bivalve larvae, and one species of pteropod larvae (*Cresis* spp.). N/A represents not available.

4. Discussion

The abundance of mesozooplankton decreased with depth and was mainly accumulated in the euphotic zone (0–200 m). In this study, the near-bottom samples were approximately 0.02% and 0.3% of the total abundance in the euphotic zone of the SF and OVF, respectively. Similar results were reported in the waters of the Guaymas Basin hydrothermal vents, where the benthopelagic zooplankton biomass was about 0.02–0.08% of the surface biomass, and the biomass of zooplankton was 10 times lower in 100–200 m of the bottom than in the surface waters [22,23]. The results of the present study demonstrated that the major accumulation of mesozooplankton was at 0–200 m in the water columns above both the hydrothermal vent fields. Among them, immature copepods of

Oithona (28%) and *Paracalanus* (16%) were dominant in SF and OVF, respectively. This may be influenced by relatively higher temperatures (>23.7 °C) at SF and higher chl-*a* concentrations in the SCM (avg. 112.5 m) at OVF. The plankton distribution was shown to differ significantly above the hydrothermal vents of the Mid-Atlantic Ridge depending on the productivity of the upper layers [26]. We studied the community structure of the water columns and near-bottom mesozooplankton because the research on the copepods among mesozooplankton in the near-bottom of hydrothermal vents has been limited. The near-bottom dominant copepods were poecilostomatoids at both the OVF (*Oncaea* and *Corycaeus*) and the SF (*Corycaeus*), and harpacticoids appeared in relatively higher abundance at near-bottom of hydrothermal vents (OVF: avg. 16%, SF: avg. 25%). Interestingly, harpacticoids did not appear at the ref-site, implying that harpacticoids may be able to feed on chemosynthetic bacteria near or above hydrothermal vents. The near-bottom copepods are able to feed on free-living chemosynthetic bacteria associated with vent effluents in areas where the effluent signature is weak, playing an important role in the transfer of vent productivity to the deep sea [41]. In addition, in chemosynthetic habitats, harpacticoids were dominant in vent epifauna and seep epifaunal communities, showing a relatively high dominance of copepods within the meiofauna communities [42]. Skebo (2004) studied copepod densities in the near-bottom water layers of the Juan de Fuca (NE Pacific) and reported densities ranging from 0.7–3.5 ind. m^{-3} over smoker vent and non-vent sites to 14.6 ind. m^{-3} over diffuse vent sites [41]. Similar results were found at the bottom plankton tows of the Guaymas Basin, in which the copepods of siphonostomes and poecilostomes (avg. 25% of the copepod population) and cyclopoids and harpacticoids were collected in all bottom plankton tows [22].

The near-bottom depth strata range at OVF (1600–1900 m) and SF (1500–2500 m) showed interesting copepod composition, distinguished from the near-bottom depth strata of hydrothermal vents. The community analysis showed that there were distinct shallow (<800 m depth) and deep (>800 m depth) faunal assemblages [24]. The distinct deep faunal assemblages at OVF included *Oncaea* copepodites (21%), foraminiferans (11%), ostracods (11%), and harpacticoid copepods (11%), and at SF, included radiolarians (22%), *Microsetella* spp. (22%), and *Oncaea* copepodites (22%). The copepods can feed directly on bacteria or dissolved organics [24]. There was a possibility of underestimation of the abundance in smaller organisms such as nauplii, copepodites, meroplanktonic larvae and copepod species of *Oncaea* and *Oithona* because of the mesh size used (200 μm) for MOCNESS in this study. According to Gallienne and Robbins (2001), an 80 μm net will collect 90% of the total zooplankton abundance and finer mesh nets may result in reduced estimates of larger taxa, and larger mesh (200 μm) are likely only to catch 7% of the total zooplankters between 200 μm and 20 μm in dimension [43].

In this study, radiolarians were the dominant zooplankton throughout the water column above the hydrothermal vent sites (OVF and SF), except at the ref-site. Vereshchaka and Vinogradov (1999) investigated zooplankton aggregations at the MAR Broken Spur, finding that both gelatinous and radiolarians dominate biomass aggregation, significantly contributing to plankton biomass throughout the water column [44]. The radiolarians were dominant near the surface (60–90% of the total biomass), in the lower part of the main pycnocline (10–30% of the total biomass), and near the plume (20–30% of the total biomass) [44]. The speculation as to why radiolarians appear throughout the water column and also near the bottom is that it may be the advantageous feeding through passive absorption of particulate matter [44].

In this study, the holoplanktonic mollusk larvae made up approximately 74%, 61%, and 92% of the total abundance of mollusk larvae in the integrated water columns of SF, OVF, and the ref-site, respectively. The holoplanktonic mollusk larvae including gastropod larval species of 1–4, 6, and 13 in this study had similar morphological features with the genus *Atlanta, Oxygrus,* and *Limacina* [38]. The holoplanktonic larvae shells are relatively flat and coiled, and the swimming fins are well developed [21]. According to Lalli and Gilmer (1989) [21], the heteropods have evolved a single swimming fin, whereas pteropods have evolved paired swimming wings, and the feeding mechanisms in open ocean differ from those in the benthic habitat due to the availability of food [21]. The holoplankton are widely distributed in the world's oceans and usually inhabit the upper ocean of generally offshore regions, where they undertake diurnal vertical migrations [21,45]. Some species of *Limacina* (ranked among the most abundant gastropods) and other abundant planktonic gastropods may have significant impacts on the epipelagic marine communities [21].

The meroplanktonic mollusk larvae including gastropod larval species of 5, 7–12, and 14–16 and bivalve larvae appeared mostly in the near-surface waters of the euphotic zone (0–200 m depth) of the study sites. Meroplankton, the planktotrophic larvae of benthic organisms, have shells with diamond shaped patterns and relatively large coils as compared with holoplanktonic larvae [27]. The planktotrophic phase is important for larval development because it allows larvae to feed during the free-swimming stage, promoting deep-sea community dispersal, colonization, and resilience [46,47]. The larvae interact with oceanic circulation, ocean ridge topography, and ridge flow and exhibit vertical migration behaviors [5,29]. Depending on the species, they spend hours to several weeks or months dispersing and drifting in the water columns by means of water currents before setting and metamorphosing on the sea floor [21].

The present study showed similar evidence that meroplanktonic mollusk larvae may disperse in the surface waters by the possibility of vertical migration. The bivalve larvae, which appeared in both the OVF and the ref-site, had morphologically similar features to planktotrophic larvae of the deep-sea hydrothermal vent species of *Bathymodiolus* spp. [27]. According to Mills et al. (2007), they collected *Bathymodiolus* species (length 400–450 μm) from near the settlements, of brownish color. *B. marisindicus* (avg. 31.1%) was one of the top-ranked microbenthic species at OVF [35]. However, we cannot be certain that the bivalve larvae are larvae of *Bathymodiolus* spp. because larvae of coastal mussels (*Mytilus* spp.) also have similar morphological features, thus further molecular analyses are needed to confirm the species [48,49]. The vent mussel *B. marisindicus* has been previously reported at the SF and OVF [34,35]. So, the bivalve larvae may have vertically migrated from the hydrothermal vents of the OVF to the surface waters. Arellano et al. (2014) reported cold methane seep species of mytilid bivalve *B. childressi* in the euphotic zone (0–100 m depth) and suggested that they had migrated to surface waters from the ocean floor [33]. The possibility of bivalve larvae being transported from the coastal, intertidal, or shallow waters maybe low because our study site is in the open ocean, approximately 1000 km away from the coasts. The dispersal of coastal, intertidal, or shallow-water bivalve larvae is through transport from inshore waters into larger scale coastal currents, and then eventually returning to the inshore waters of another location [9].

The gastropod larvae (species 7) that appeared mostly in the euphotic zone (0–200 m) at OVF and SF had morphological similarities to gastropod larvae Lepetodrilidae (*Lepetodrilus* spp.), which inhabit the hydrothermal vent sites of Juan de Fuca Ridge and EPR [27,44]. The abundance of vent gastropod larvae (*Lepetodrilus* sp. and two peltospirid species) is significantly higher in the plume than away from it (mean abundance = 21.0 ind. 1000 m^{-3}) [50]. Gastropod larval species 5, 8, 12, 14, and 15 have similar morphological features to macrobenthic gastropod larvae of *Vetulonia* spp. that were found in the hydrothermal vents and hydrothermal seeps at the North Fiji Basin [51]. Gastropod larval species 9–11 and 16 have morphological features similar to those of *Phymorphynchus protoconchs* [27]. The vent mollusk fauna in SF includes *B. marisindicus* (Mitilidae), *Lepetodrilus* sp. (Lepetodridae), *Eulepetopsis* sp. (Neolepetopsidae), Scaly-foot gastropod (Peltospiridae), *Shinkailepas* spp. (Phenacolepadidae), *Alviniconcha* spp., *Desbruyeresia* spp. (Provannidae), and *Phymorhynchus* spp. (Turridae) [34]. The vent mollusk fauna in OVF includes *B. marisindicus*, *B.* sp.1, *Gigantidas* n. spp., *Alvinocaris markensis*, *Paralepetopsis ferrugivora*, and *Lepetodrilus* spp. [35]. Larval dispersal in the planktonic larval stage is an important component of the life cycle of species, particularly in isolated habitats such as hydrothermal vents [28]. Adams et al. (2012) studied and reviewed the reproductive patterns and larval durations, behaviors and vertical distributions, and the coupling of vent topography and hydrodynamics that affect dispersal [5]. Yahagi et al. (2017) provided evidence of early life history traits and population genetics for the surface dispersal of gastropod vent species *Shinkailepas myojinensis* that inhabited sulfide chimneys and volcanic rocks covered by bacterial mats in diffuse venting areas [6]. The hatched larvae swam upwards at 16.6–44.42 mm/min depending on temperature, and were likely to migrate to surface water because of the high food source (high phytoplankton biomass) that helps in their growth and development [6]. Therefore, in this study, the gastropod larvae in the surface waters may have dispersed or vertically migrated from the deep-sea hydrothermal vent sites (OVF and SF).

5. Conclusions

The quantitative measurements of mesozooplankton including holoplankton and meroplankton were conducted in the OVF, SF, and ref-site. The integrated abundance of mesozooplankton over the water column was relatively higher in the ref-site (77,037 ind. 100 m^{-3}) than in the vent sites (avg. 53,398 ind. 100 m^{-3}). It is evident that most of the mesozooplankton was accumulated in the euphotic zone (0–200 m). Mollusk larvae (gastropods and bivalves) appeared at all stations. Only one species of bivalve larva that appeared at the OVF and ref-site (0–200 m) was morphologically similar to *Bathymodiolus* spp. that inhabits the hydrothermal vents. The gastropod larvae were morphologically identified as holoplanktonic and meroplanktonic. Sixteen species of gastropod larvae were distinguished morphologically. Gastropod larvae species 1–4, 6, and 13 had holoplanktonic features (*Atlanta, Oxygrus,* and *Limacina*), while gastropod species 5, 7–12, and 14–16 had meroplanktonic features. Gastropod larvae species 5, 11, and 12 only appeared at OVF, 9 and 10 only appeared at SF, 4–16 appeared only at ref-site, and gastropod larvae species 7–8 appeared in all surveyed stations. Gastropod larvae species 5, 8, 12, 14, and 15 have similar morphological features to the macrobenthic gastropod larvae of *Vetulonia* spp. Gastropod larvae species 7 have morphological features similar to gastropods Lepetodrilidae, and species 9–11 and 16 have morphological features similar to *Phymorphynchus protoconchs*. The morphologically identified mollusk larvae of gastropods and bivalves in the upper layers of the water column (0–200 m) indicate that larvae associated with the hydrothermal vents may disperse approximately 2000 m above the vents.

Author Contributions: Conceptualization, M.K. and J.-H.K.; methodology, M.K. and J.-H.K.; investigation, M.K., J.-H.K. and D.K.; data curation, M.K., J.-H.K.; writing—original draft preparation, M.K., J.-H.K.; writing—review and editing, M.K., J.-H.K.; visualization, M.K., J.-H.K.; project administration, M.K., J.-H.K. and D.K.; funding acquisition, J.-H.K. and D.K. All authors have read and agreed to the published version of the manuscript.

Funding: This research was a part of the project titled "Understanding the deep-sea biosphere on seafloor hydrothermal vents in the Indian Ridge (No. 20170411)" funded by the Ministry of Oceans and Fisheries, Korea.

Institutional Review Board Statement: Not applicable.

Informed Consent Statement: Not applicable.

Data Availability Statement: All data used are presented in the text, figures, and tables.

Acknowledgments: The authors would like to thank the captain and the crews of R/V ISABU for supporting the MOCNESS operations and shipboard works. We thank G. Hwang for bathymetry assistance.

Conflicts of Interest: The authors declare no conflict of interest.

References

1. Lutz, R.A.; Jablonski, D.; Turner, R.D. Larval development and dispersal at deep-sea hydrothermal vents. *Science* **1984**, *226*, 1451–1454. [CrossRef] [PubMed]
2. Bouchet, P.; Warén, A. Ontogenetic migration and dispersal of deep-sea gastropod larvae. In *Reproduction, Larval Biology, and Recruitment of the Deep-Sea Benthos*; Young, C.M., Eckelbarger, K.J., Eds.; Columbia University Press: New York, NY, USA, 1994; pp. 98–117.
3. Keough, M.J.; Swearer, S.E. Early life histories of marine invertebrates and fishes. In *Marine Ecology*; Connell, S.D., Gillanders, B.M., Eds.; Oxford University Press: Oxford, UK, 2007; pp. 19–46.

4. Swearer, S.E.; Treml, E.A.; Shima, J.S. A review of biophysical models of marine larval dispersal. In *Oceanography and Marine Biology*; Hawkins, S.J., Allcock, A.L., Bates, A.E., Firth, L.B., Smith, I.P., Swearer, S.E., Todd, P.A., Eds.; CRC Press: Boca Raton, FL, USA, 2019; Volume 57, pp. 325–356.
5. Adams, D.K.; Arellano, S.M.; Govenar, B. Larval dispersal: Vent life in the water column. *Oceanography* **2012**, *25*, 256–268. [CrossRef]
6. Yahagi, T.; Kayama Watanabe, H.; Kojima, S.; Kano, Y. Do larvae from deep-sea hydrothermal vents disperse in surface waters? *Ecology* **2017**, *98*, 1524–1534. [CrossRef] [PubMed]
7. Young, C.M.; Arellano, S.M.; Hamel, J.F.; Mercier, A. Ecology and evolution of larval dispersal in the deep sea. In *Evolutionary and Ecology of Marine Invertebrate Larvae*; Carrier, T.J., Reitzel, A.M., Heyland, A., Eds.; Oxford University Press: Oxford, UK, 2018; pp. 229–245.
8. Shanks, A.L. Pelagic larval duration and dispersal distance revisited. *Biol. Bull.* **2009**, *21*, 373–385. [CrossRef]
9. Tilburg, C.E.; McCartney, M.A.; Yund, P.O. Across-shelf transport of bivalve larvae: Can the interface between a coastal current and inshore waters act as an ecological barrier to larval dispersal? *PLoS ONE* **2012**, *7*, e48960. [CrossRef]
10. Mullineaux, L.S.; Mills, S.W.; Sweetman, A.K.; Beaudreau, A.H.; Metaxas, A.; Hunt, H.L. Vertical, lateral and temporal structure in larval distributions at hydrothermal vents. *Mar. Ecol. Prog. Ser.* **2005**, *293*, 1–16. [CrossRef]
11. Brooke, S.D.; Young, C.M. Where do the embryos of *Riftia pachyptila* develop? Pressure tolerances, temperature tolerances, and buoyancy during prolonged embryonic dispersal. *Deep-Sea Res. Part II* **2009**, *56*, 1599–1606. [CrossRef]
12. Arellano, S.M.; Young, C.M. Temperature and salinity tolerances of embryos and larvae of the deep-sea mytilid mussel *"Bathymodiolus" childressi*. *Mar. Biol.* **2011**, *158*, 2481–2493. [CrossRef]
13. Nakamura, K.; Watanabe, H.; Miyazaki, J.; Takai, K.; Kawagucci, S.; Noguchi, T.; Okamura, K. Discovery of new hydrothermal activity and chemosynthetic fauna on the Central Indian Ridge at 18°–20°S. *PLoS ONE* **2012**, *7*, e32965. [CrossRef]
14. Lonsdale, P. Clustering of suspension-feeding macrobenthos near abyssal hydrothermal vents at oceanic spreading centers. *Deep Sea Res.* **1977**, *24*, 857–863. [CrossRef]
15. Nakamura, K.; Takai, K. Indian Ocean Hydrothermal Systems: Seafloor Hydrothermal Activities, Physical and Chemical Characteristics of Hydrothermal Fluids, and Vent-Associated Biological Communities. In *Subseafloor Biosphere Linked to Hydrothermal Systems*; Ishibashi, J.-I., Okino, K., Sunamura, M., Eds.; Springer: Tokyo, Japan, 2015; pp. 147–161.
16. Martin, W.; Baross, J.; Kelley, D.; Russell, M.J. Hydrothermal vents and the origin of life. *Nat. Rev. Microbiol.* **2008**, *6*, 805–814. [CrossRef]
17. Demina, L.L.; Galkin, S.V. *Trace Metal Biogeochemistry and Ecology of Deep-Sea Hydrothermal Vent Systems*, 1st ed.; Springer: Cham, Switzerland, 2016; Volume 50, pp. 16–17.
18. Van Dover, C.L.; Arnaud-Haond, S.; Gianni, M.; Helmreich, S.; Huber, J.A.; Jaeckel, A.L.; Metaxas, A.; Pendleton, L.H.; Petersen, S.; Ramirez-Llodra, E.; et al. Scientific rationale and international obligations for protection of active hydrothermal vent ecosystems from deep-sea mining. *Mar. Policy* **2018**, *90*, 20–28. [CrossRef]
19. Trivett, D.A.; Williams, A.J., III. Effluent from diffuse hydrothermal venting: 2. Measurement of plumes from diffuse hydrothermal vents at the southern Juan de Fuca Ridge. *J. Geophys. Res. Oceans* **1994**, *99*, 18417–18432. [CrossRef]
20. Beedessee, G.; Watanabe, H.; Ogura, T.; Nemoto, S.; Yahagi, T.; Nakagawa, S.; Nakamura, K.; Takai, K.; Koonjul, M.; Marie, D.E.; et al. High connectivity of animal populations in deep-sea hydrothermal vent fields in the Central Indian Ridge relevant to its geological setting. *PLoS ONE* **2013**, *8*, e81570. [CrossRef]
21. Lalli, C.M.; Gilmer, R.W. *Pelagic Snails: The Biology of Holoplanktonic Gastropod Mollusks*, 1st ed.; Stanford University Press: Redwood City, CA, USA, 1989; pp. 1–6. ISBN 9780804714907.
22. Berg, C.J., Jr.; Van Dover, C.L. Benthopelagic macrozooplankton communities at and near deep-sea hydrothermal vents in the eastern Pacific Ocean and the Gulf of California. *Deep Sea Res. Part A Oceanogr. Res. Pap.* **1987**, *34*, 379–401. [CrossRef]
23. Wiebe, P.H.; Copley, N.; Van Dover, C.; Tamse, A.; Manrique, F. Deep-water zooplankton of the Guaymas Basin hydrothermal vent field. *Deep Sea Res. Part I Oceanogr. Res. Pap.* **1988**, *35*, 985–1013. [CrossRef]
24. Burd, B.J.; Thomson, R.E. Hydrothermal venting at Endeavour Ridge: Effect on zooplankton biomass throughout the water column. *Deep Sea Res. Part A Oceanogr. Res. Pap.* **1994**, *41*, 1407–1423. [CrossRef]
25. Burd, B.J.; Thomson, R.E. Distribution of zooplankton associated with the Endeavour Ridge hydrothermal plume. *J. Plankton Res.* **1995**, *17*, 965–997. [CrossRef]
26. Vinogradov, G.M.; Vereshchaka, A.L.; Aleinik, D.L. Zooplankton distribution over hydrothermal fields of the Mid-Atlantic Ridge. *Oceanol. C/C Okeanol.* **2003**, *43*, 656–669.
27. Mills, S.W.; Beaulieu, S.E.; Mullineaux, L.S. Photographic Identification Guide to Larvae at Hydrothermal Vents in the Eastern Pacific. 2009. Available online: http://www.whoi.edu/science/B/vent-larval-id (accessed on 13 December 2021).
28. Kim, S.L.; Mullineaux, L.S.; Helfrich, K.R. Larval dispersal via entrainment into hydrothermal vent plumes. *J. Geophys. Res. Oceans* **1994**, *99*, 12655–12665. [CrossRef]
29. Adams, D.K.; Mullineaux, L.S. Supply of gastropod larvae to hydrothermal vents reflects transport from local larval sources. *Limnol. Oceanogr.* **2008**, *53*, 1945–1955. [CrossRef]
30. Kim, S.L.; Mullineaux, L.S. Distribution and near-bottom transport of larvae and other plankton at hydrothermal vents. *Deep Sea Res. Part II Top. Stud. Oceanogr.* **1998**, *45*, 423–440. [CrossRef]

31. Adams, D.K.; Mills, S.W.; Shank, T.M.; Mullineaux, L.S. Expanding dispersal studies at hydrothermal vents through species identification of cryptic larval forms. *Mar. Biol.* **2010**, *157*, 1049–1062. [CrossRef]
32. Adams, D.K.; McGillicuddy, D.J.; Zamudio, L.; Thurnherr, A.M.; Liang, X.; Rouxel, O.; German, C.R.; Mullineaux, L.S. Surface-generated mesoscale eddies transport deep-sea products from hydrothermal vents. *Science* **2011**, *332*, 580–583. [CrossRef]
33. Arellano, S.M.; Van Gaest, A.L.; Johnson, S.B.; Vrijenhoek, R.C.; Young, C.M. Larvae from deep-sea methane seeps disperse in surface waters. *Proc. R. Soc. B Biol. Sci.* **2014**, *281*, 20133276. [CrossRef]
34. Watanabe, H.; Beedessee, G. Vent fauna on the Central Indian Ridge. In *Subseafloor Biosphere Linked to Hydrothermal Systems*; Ishibashi, J.-I., Okino, K., Sunamura, M., Eds.; Springer: Tokyo, Japan, 2015; pp. 205–212.
35. Kim, J.; Son, S.K.; Kim, D.; Pak, S.J.; Yu, O.H.; Walker, S.L.; Oh, J.; Choi, S.K.; Ra, K.; Ko, Y.; et al. Discovery of active hydrothermal vent fields along the Central Indian Ridge, 8–12°S. *Geochem. Geophys. Geosyst.* **2020**, *21*, e2020GC009058. [CrossRef]
36. Hood, R.R.; Beckley, L.E.; Wiggert, J.D. Biogeochemical and ecological impacts of boundary currents in the Indian Ocean. *Prog. Oceanogr.* **2017**, *156*, 290–325. [CrossRef]
37. Parsons, T.R.; Maita, Y.; Lalli, C.M. *A Manual of Chemical and Biological Methods for Seawater Analysis*; Pergamon Press: New York, NY, USA, 1984; pp. 101–112.
38. Conway, D.V.; White, R.G.; Hugues-Dit-Ciles, J.; Gallienne, C.P.; Robins, D.B. *Guide to the Coastal and Surface Zooplankton of the South-Western Indian Ocean. DEFRA Darwin Initiative Zooplankton Programme, Marine Biological Association of the United Kingdom, No. 15*; Marine Biological Association of the United Kingdom: Plymouth, UK, 2003; pp. 1–354. ISSN 0260-2784.
39. Chihara, M.; Murano, M. *An Illustrated Guide to Marine Plankton in Japan*; Tokai University Press: Hiratsuka, Japan, 1997.
40. Kenk, V.C.; Wilson, B.R. A new mussel (Bivalvia, Mytilidae) from hydrothermal vents in the Galapagos Rift zone. *Malacologia* **1985**, *26*, 253–271.
41. Skebo, K.M. Distribution of zooplankton and nekton above hydrothermal vents on the Juan de Fuca and Explorer ridges. Master's Thesis, University of Victoria, Victoria, BC, Canada, 2004.
42. Gollner, S.; Ivanenko, V.N.; Arbizu, P.M.; Bright, M. Advances in taxonomy, ecology, and biogeography of Dirivultidae (Copepoda) associated with chemosynthetic environments in the deep sea. *PLoS ONE* **2010**, *5*, e9801. [CrossRef] [PubMed]
43. Gallienne, C.P.; Robins, D.B. Is *Oithona* the most important copepod in the world's oceans? *J. Plankton Res.* **2001**, *23*, 1421–1432. [CrossRef]
44. Vereshchaka, A.L.; Vinogradov, G.M. Visual observations of the vertical distribution of plankton throughout the water column above Broken Spur vent field, Mid-Atlantic Ridge. *Deep Sea Res. Part I Oceanogr. Res. Pap.* **1999**, *46*, 1615–1632. [CrossRef]
45. Hart, M.B.; Wall-Palmer, D.; Janssen, A.W.; Smart, C.W. Some observations on the geological history of the holoplanktonic gastropods. *Proc. Geol. Assoc.* **2020**, *131*, 443–449. [CrossRef]
46. Lutz, R.A.; Kennish, M.J. Ecology of deep-sea hydrothermal vent communities: A review. *Rev. Geophys.* **1993**, *31*, 211–242. [CrossRef]
47. Baco, A.R.; Etter, R.J.; Ribeiro, P.A.; Von der Heyden, S.; Beerli, P.; Kinlan, B.P. A synthesis of genetic connectivity in deep-sea fauna and implications for marine reserve design. *Mol. Ecol.* **2016**, *25*, 3276–3298. [CrossRef] [PubMed]
48. Sprung, M. Physiological energetics of mussel larvae (*Mytilus edulis*). I. Shell growth and biomass. *Mar. Ecol. Prog. Ser.* **1984**, *17*, 283–293. [CrossRef]
49. Schönitzer, V.; Weiss, I.M. The structure of mollusc larval shells formed in the presence of the chitin synthase inhibitor Nikkomycin Z. *BMC Struct. Biol.* **2007**, *7*, 71. [CrossRef]
50. Mullineaux, L.S.; Wiebe, P.H.; Baker, E.T. Larvae of benthic invertebrates in hydrothermal vent plumes over Juan de Fuca Ridge. *Mar. Biol.* **1995**, *122*, 585–596. [CrossRef]
51. Warèn, A.; Bouchet, P. New records, species, genera, and a new family of gastropods from hydrothermal vents and hydrocarbon seeps. *Zoologica Scripta* **1993**, *22*, 1–90. [CrossRef]

Journal of
Marine Science and Engineering

MDPI

Article

New Record of Hydrothermal Vent Squat Lobster (*Munidopsis lauensis*) Provides Evidence of a Dispersal Corridor between the Pacific and Indian Oceans

Hee-seung Hwang [1], Boongho Cho [2,3], Jaemin Cho [3], Beomseok Park [4,*] and Taewon Kim [2,3,*]

[1] Research Institute of EcoScience, Ewha Womans University, Seoul 03760, Korea; winsome212@nate.com
[2] Program in Biomedical Science and Engineering, Inha University, 100 Inha-ro, Incheon 22212, Korea; boonghocho@gmail.com
[3] Department of Ocean Sciences, Inha University, 100 Inha-ro, Incheon 22212, Korea; cjm970131@naver.com
[4] Department of Biomedical Laboratory Science, College of Health Science, Eulji University, Seongnam 13135, Korea
* Correspondence: bspark74@eulji.ac.kr (B.P.); ktwon@inha.ac.kr (T.K.); Tel.: +82-10-8726-3070 (T.K.)

Abstract: Hydrothermal vents are chemosynthetically driven ecosystems and one of the most extreme environments on Earth. Vent communities exhibit remarkable taxonomic novelty at the species and supra-species levels, and over 80% of vent species are endemic. Here, we used mitochondrial DNA to identify the biogeographic distribution of *Munidopsis lauensis* and the heme-binding regions of A1-type COX1 from six species (including *M. lauensis*) to investigate whether genetic variation in the protein structure affects oxygen-binding ability. We verified the identity of Indian Ocean specimens by comparing sequences from the barcoding gene mitochondrial cytochrome oxidase subunit 1 (COI) with known *M. lauensis* sequences from the NCBI database. The data show that these are the first recorded specimens of *M. lauensis* in the Indian Ocean; previously, this species had been reported only in the southwest Pacific. Our findings support the hypothesis that vent fauna in the Pacific and Indian Oceans can interact via active ridges. In the case of the mitochondrial DNA-binding site, the arrangement of heme-binding ligands and type A1 motif of *M. lauensis* was identical to that in other species. Moreover, our findings suggest that the mechanism of oxygen binding is well conserved among species from terrestrial organisms to hydrothermal extremophiles. Overall, dispersal of the same species to geologically separated hydrothermal vents and conserved heme-binding regions in mitochondrial proteins suggest that hydrothermal species might have evolved from shallow sea organisms and became distributed geographically using a dispersion corridor.

Keywords: hydrothermal vents; Munidopsis; dispersal corridor; vent fauna; heme-binding site

1. Introduction

Chemosynthetically driven hydrothermal vents are among the harshest ecosystems on the planet, having temperatures as high as 390 °C, low oxygen levels, and enriched concentrations of hydrogen sulfide (H_2S), methane (CH_4), and heavy metals, including iron, zinc, and copper [1]. This unique environment makes vent species susceptible to geological settings and local ecosystems because they produce chemical-rich fluids that support the food web, generating a patchwork seafloor habitat. Catastrophic disturbances to vent species or vents can eradicate entire communities [2]. The characteristics of vent communities are complex, and although 82% of vent species are endemic, they are remarkably diverse at multiple taxonomy levels (e.g., family, order, class, and species) [3].

Hydrothermal vents are found at mid-ocean spreading centers in the Atlantic, Arctic, Indian, and eastern Pacific Oceans, and back-arc basins of the western Pacific Ocean [4]. In particular, numerous vent communities have been reported along active margins in the Atlantic [5,6], Pacific [7,8], and Indian Oceans [9,10].

Phylogeography is a critical ecological characteristic that explains a species' evolutionary history and successful adaptation to environmental changes. In terms of community, most organisms in Indian Ocean vent fields have evolutionary relationships with western Pacific vent fauna [10]. However, exceptions exist, such as the shrimp *Rimicaris aff. exoculata*, a decapod that is the predominant species in Indian Ocean vents but is similar to its mid-Atlantic counterpart [10]. Global assessments of chemosynthetic faunal biogeography have suggested that the Indian Ocean vent communities follow asymmetric assembly rules biased toward Pacific evolutionary alliances [10].

Decapods (e.g., alvinocaridid shrimp, bythograeid crabs, and galatheid squat lobsters) represent approximately 10% of taxa in hydrothermal vents and are the dominant fauna [1,11,12]. Genetics appear to have participated in bythograeid crab and alvinocaridid shrimp adaptation to vent habitats [13,14]. Munidopsis is the second-largest and ecologically diverse genus of galatheid squat lobsters. Globally, more than 200 species have been defined: more than 150 species in the Indo-Pacific and at least 70 in the Atlantic [15,16]. Although only 10 species are endemic to hydrothermal vent environments [17–22], they exhibit a unique pattern of distribution and abundance. Five species are found in the western Pacific (*M. starmer*, *M. sonne*, *M. lauensis*, *M. marianica*, and *M. myojinensis*), three in the eastern Pacific (*Munidopsis* sp., *M. subsquamosa*, and *M. lentigol*), and two in the Mid-Atlantic Ridge (*M. exuta* and *M. acutispina*) [22]. Interestingly, all Munidopsis species have limited distributions, comprising three coexisting species at most (Table 1).

Table 1. Diversity of *Munidopsis* in hydrothermal vent.

Ocean	Hydrothermal Vent	Depth	Species	Reference
Western Pacific Ocean	Brother Seamount (34°51′45.4″ S, 179°03′28.6″ E)	1649–1750 m 1649–1992 m 1649 m	*M. lauensis* *M. sonne* *M. kermadec*	[23]
	North Fiji Basin (16°59′00.0″ S, 173°55′00.0″ E)	no record	*M. lauensis* *M. sonne* *M. starmer*	[21]
	Lau Basin (20°59′21.0″ S, 176°34′06.0″ W)	1750 m	*M. lauensis*	[21,24]
	Izu-Bonin arc (32°06.25′ N, 139°52.17′E)	1288–1625 m	*M. myojinensis*	[21]
	Manus Basin (3°43′49.4″ S, 151°40′27.5″ E)	no record	*M. lauensis*	[21]
	Mariana (18°11′00.0″ N, 144°45′00.0″ E)	no record	*M. marianica*	[21]
	Formosa Ridge (22°06′54.0″ N 119°17′06.0″ E)	1750–2000 m	*M. lauensis*	[25]
Indian Ocean	**Onnuri Vent Field(11°24′54.9″ S, 66°25′24.0″ E)**	**2023 m**	***M. lauensis***	**the present study**
	Kairei Field (25°19′14.4″ S, 70°02′24.0″ E)	2422–2435 m	*M. larticorpus*	[22]
	Forecast Vent Field/Mariana Back Arc Basin (13°23.07′ N 145°55.02′ E)	1450 m	*M. gracilis*	[22]
Mid-Atlantic Ridge	Mid-Atlantic Ridge (1°38′50.7″ N 19°42′41.1″ W)	no record	*M. acutispina*	[22]
	Mid-Atlantic Ridge (2°18′10.9″ N 25°38′33.9″ W)	no record	*M. exuta*	[21,26]
	East Pacific Rise S of Baja California (20°49′36.0″ N 109°06′00.0″ W)	3502 m	*M. lentigo*	
Northern Pacific Ocean	Galapagos Rift (eastern Pacific 13° N and 21° N areas)	no record	*M. subsquamosa*	[25]
	Limbo Vent, Juan De Fuca Ridge (46°00′02.0″ N 129°59′59.1″ W)	1545–2008 m	*M. alvisca*	[17]

Not all vent fauna have distributions as restricted as Munidopsis. Molecular comparisons have revealed genetic affinity between the following species in the Indian and western Pacific Oceans: shrimp *Rimicaris kairei*, gastropod Alviniconcha sp., bythograeid crab *Austinograea rodriguezensis*, stalked barnacle *Neolepas* sp., deep-sea mussel *Bathymodiolus marisindicus*, sea anemone Marianactis sp., and scaly foot gastropod [10,27].

Mitochondria are responsible for most cellular aerobic metabolism, producing ATP through the electron transport chain. All 13 mitochondrial protein-coding genes are involved in this process. Functional restrictions on mitochondrial genes are related to the following adaptive evolutionary mechanisms: climatic adaptation [28], locomotion [28,29], high elevation adaptation (low-oxygen and cold climate) [30–35], mammalian adaptation [36], and deep-sea hydrostatic pressure adaptation [37]. DNA barcoding with a cytochrome c oxidase subunit 1 (COX1) region has been widely used to recognize species in taxonomic studies [38]. The COX1 region is divided into three evolutionary families: type A (mitochondrial-like oxidases), B (ba_3-like oxidases), and C (cbb_3-type oxidases). The structural diversity of these families correlates with different proton pumping efficiencies [39].

In this study, we investigated the biogeographic distribution of *Munidopsis lauensis* and tested a hypothesis that the Indian Ocean is a dispersal corridor connecting the hydrothermal vent fauna of the Atlantic and Pacific Oceans. We also compared the heme-binding regions of proteins from six species, including *M. lauensis*, to evaluate whether genetic variation affects oxygen-binding ability. Overall, the purpose of this study was to obtain information about the evolutionary history of *M. lauensis*, including its dispersal mechanisms, using mitochondrial data.

2. Materials and Methods

2.1. Sampling and Mapping

On 1 July 2019, eighteen samples of *M. lauensis* were collected at the Onnuri Vent Field (OVF; latitude: 11°14′58.6″ S; longitude: 66°15′14.4″ E) in the Indian Ocean using a TeleVision-grab (TV-grab) device mounted on the Research Vessel (R/V) ISABU (Dive No. 6). Samples were immediately fixed in 75% ethanol (Figure 1). The sampling depth was approximately 2023.2 m.

Figure 1. *Munidopsis lauensis* Baba and de Saint Laurent, 1992.

A map (Figure 2A) was developed using a Diva-Gis 7.5 template (http://www.diva-gis.org/ accessed on 3 January 2022). Sampling locations of the specimens included in the analyses and data from Major Ocean Currents (2016) and Global Distribution of Hydrothermal Vent Fields (2020) in ArcGIS were combined.

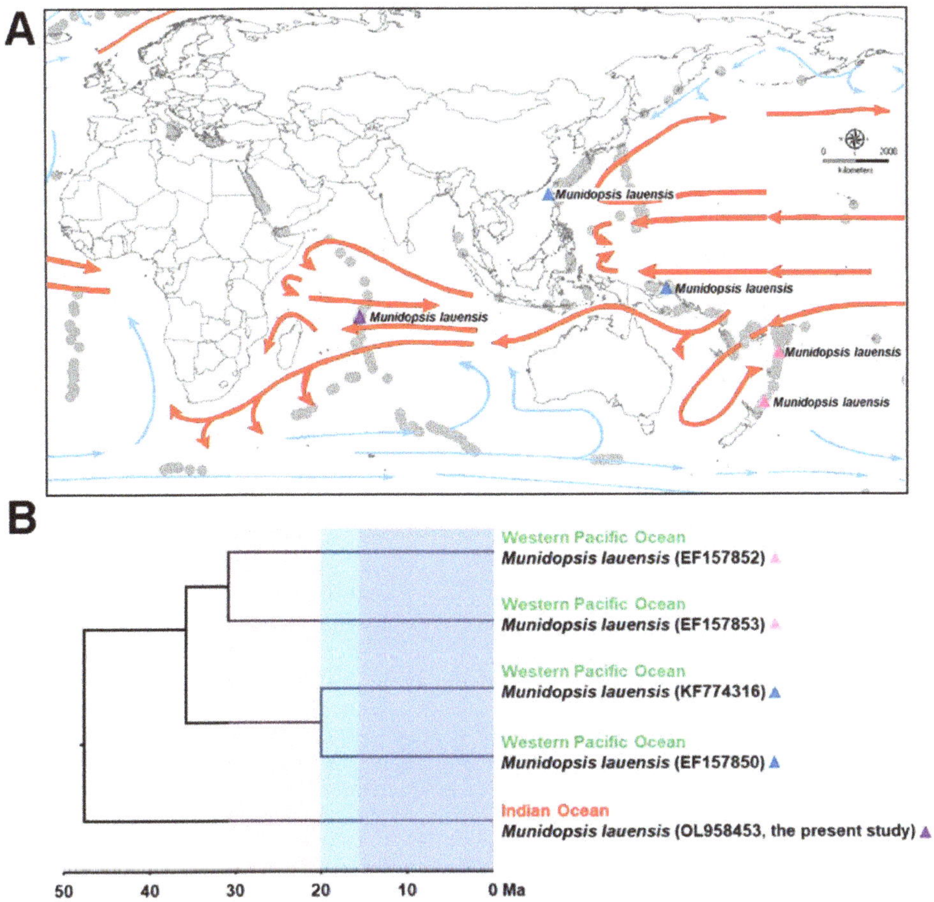

Figure 2. (**A**) Biogeographic distribution of *M. lauensis*. Red arrows indicate warm currents and blue arrows indicate cold currents. (**B**) Schematic diagram of divergence analysis for *M. lauensis*.

2.2. DNA Barcoding

Genomic DNA was extracted from the muscle tissue of five *M. lauensis* specimens using a Qiagen DNeasy Blood and Tissue kit. The barcoding regions of the mitochondrial cytochrome oxidase subunit 1 (COI) genes (658 bp) were sequenced and compared with published sequences from western Pacific Ocean isolates (accession numbers: EF157850–EF157853) and *M. verrilli* Benedict, 1902. New sequences from Indian Ocean specimens were registered in GenBank (accession numbers: OL958453–OL958457). *Shinkaia crosnieri* Baba and Williams, 1998 (accession numbers: MK795354–MK795356) were selected as closely related members of Munidopsidae based on the complete mitochondrial genome [40]. *Munida gregaria* (Fabricius 1793) (accession numbers: NC030255, KU521508) was selected because Mundidae and Mundopsidae are phylogenetically related families [40]. *Munida gregaria* and *Shinkaia crosnieri* were used as outgroups.

Primers used to amplify COI-5 by PCR were LCO1490 (5′-GGT CAA ATC ATA AAG ATA TTG G-3′) and HCO2198 (5′-TAA ACT TCA GGG TGA CCA AAA AAT CA-3′) [41]. The following thermocycling program was used: 5 min at 94 °C; 40 cycles of 1 min at 94 °C, 1 min at 40 °C, and 2 min at 72 °C; and a final extension at 72 °C for 10 min. The 25 μL reaction mix included 15.7 μL ultrapure water, 5 μL of 5X PCR buffer, 2 μL of each primer (10 μM), 1 μL of dNTP (10 mM), 0.3 μL of Taq polymerase (5 U), and 1 μL of DNA template. Sequences were aligned using MAFFT [42]. Sequence divergence between individuals was quantified using the Kimura 2-parameter (K2P) distance model [43]. A neighbor-joining (NJ) tree of K2P distances was created in MEGA X [44].

2.3. Estimation of Divergence Time

The most appropriate model to estimate divergence time was the HKY+Gamma model, which was selected using PartitionFinder version 1.1.1, Australia [45]. The Bayesian phylogenetic software BEAST version 2, USA [46] was used to estimate the divergence time. Analyses were performed using a strict clock and an uncorrelated lognormal relaxed molecular clock to check for rate variation among branches [47]. We conducted analyses using a Yule speciation model and a birth-death model for the tree prior to evaluate whether the sensitivity of the results was affected by the choice of tree prior. Posterior distributions of parameters were estimated using Markov chain Monte Carlo (MCMC) sampling over 10^8 steps, with samples drawn every 10^4 steps. The initial 10% of samples were discarded as burn-in. Convergence was checked by running the analysis in duplicate and visualizing the results in the program Tracer version version 1.7.1, USA [48], which showed that the effective sample size of all parameters was above 200. We used Tree Annotator version 1.8.4, USA [49], available in the BEAST version 2, USA package, to identify the maximum clade credibility tree.

2.4. Heme-Binding Site of COX1 Sequences

We used COX 1 sequences from five species classified as "type of COX1" to analyze the heme-binding region in *M. lauensis* sequences. Sequences from *Homo sapiens, Bos taurus, Escherichia coli* bo3, *Drosophila melanogaster*, and *Portunus trituberculatus* were collected from NCBI (https://www.ncbi.nlm.nih.gov/ accessed on 3 January 2022) and compared with the *M. lauensis* data. The heme-binding site sequences were aligned by multiple sequence alignment using Clustal Omega (https://www.ebi.ac.uk/Tools/msa/clustalo/ accessed on 3 January 2022) [50]. In this comparison, we focused on whether each species contains the same region found in human COX1. Secondary structures based on that of human COX1 were analyzed by comparing amino acids using ESPript 3.0 (https://espript.ibcp.fr/ESPript/ESPript/ accessed on 3 January 2022) [51]. The model structure of *M. lauensis* COX1 was calculated using a SWISS-MODEL (https://swissmodel.expasy.org/ accessed on 3 January 2022) [52]. The structural data of Human COX1 (PDB code: 5z62) and *M. lauensis* COX1 were described in PyMOL (https://pymol.org/2/ accessed on 3 January 2022) [53]. The superposition showed that the two proteins have secondary structure alignment and helped to identify heme-binding sites and type A1 motifs in *M. lauensis* COX1.

3. Results and Discussion

3.1. Biogeographic Distribution of M. lauensis

Indian Ocean *M. lauensis* specimens (accession numbers: OL958453–OL958457) have COI-barcoding regions nearly identical to West Pacific specimens (Brothers Seamount, Manus, and Lau Basins) (accession numbers: EF157850, 1587852, and 157853) [21], with a maximum of 0.58% sequence divergence (Figure 3). We conclude that our collected specimens are indeed *M. lauensis* and represent the first records of this species in the Indian Ocean. All other isolates of *M. lauensis* were from deep-sea hydrothermal vents in the southwest Pacific Ocean [54].

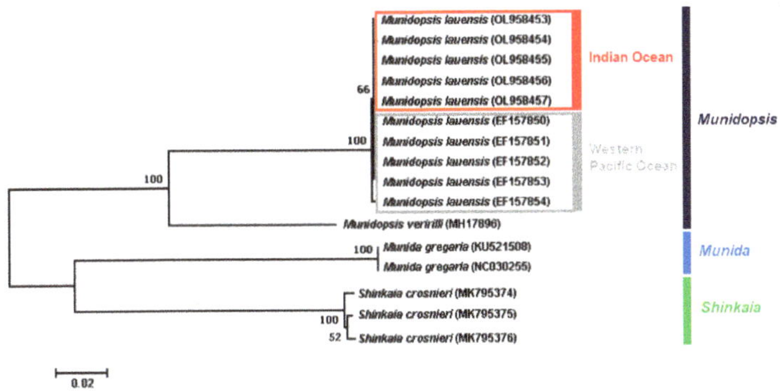

Figure 3. Genetic relationships between *Munidopsis lauensis* isolates from the Indian and western Pacific Oceans based on mitochondrial cytochrome oxidase subunit 1 (COI) sequences. *Munida gregaria* and *Shinkaia crosnieri* were used as outgroups.

The presence of this species in different oceans supports the hypothesis that the Indian Ocean is a dispersal corridor for hydrothermal vent fauna between the Atlantic and Pacific Oceans. Indeed, other faunal and molecular comparisons have also revealed an affinity between taxa in the Indian and western Pacific Oceans [10] (Figure 2A).

The order of divergence times was as follows: First, among the western Pacific *M. lauensis* isolates, the sequences from the Hine Hina material (accession number: EF157852) and the Brothers Seamount material (GenBank number: EF157853) were nearly identical and first diverged at 30 Myr. Second, the sequences from the Desmos material (accession number: EF157850) and Taiwan (accession number: KF774316) diverged at 20 Myr. Lastly, our material from the Indian Ocean (accession numbers: OL958453–OL958454) diverged at approximately 15 Myr (Figure 2B). Although it is impossible to review all distributions of the present species *M. lauensis*, this study reveals that the western Pacific group diverged before the Indian Ocean group.

Our study is the first to suggest that a dispersal corridor exists between the western Pacific and Indian Oceans, based on the occurrence of *M. lauensis* in both regions. Furthermore, our data are aligned with a previous study demonstrating the existence of a dispersal corridor between the Pacific and Atlantic Oceans [23] and therefore supports the hypothesis that dispersal corridors for hydrothermal vent species can reside in different oceans. Larvae of hydrothermal vent species floating on the ocean surface might have spread on ocean currents and then have settled onto various hydrothermal vents [55]. Although the precise dispersal mechanism is unknown, even if the probability that floating larvae could successfully settle in different locations is low, it could still have happened.

3.2. Binding Sites of COX in M. lauensis

COI proteins are categorized into three evolutionary families: Type A (mitochondrial-like oxidases), B (ba3-like oxidases), and C (cbb3-type oxidases). The structural diversity of COX1 reflects differences in proton pumping efficiency [39]. The protein sequence of type A COX1 comprises the motif -GHPEVY-. The helix 6 residues divide this motif into two subfamilies: type A1 (glutamate residue in the motif -XGHPEV-) and type A2 (tyrosine and serine in the alternative motif -YSHPXV-). The type A1 motifs of *M. lauensis* and other species are homologous (Figure 4). The heme-binding sites of human COX1 are known as H61, H240, H290, H291, H376, H378, and Y244 [56]. Seven heme-binding amino acids are arranged on the same helix in all species. This analysis shows that even if organisms have evolved in different environments, the arrangements of heme-binding ligands and type A1 motifs remain unchanged in *M. lauensis* and other species. Type A1 COX1 regions are highly conserved, having practically identical structures across numerous species (Figure 5).

However, proton transport efficiency can vary among organisms. More research is needed to evaluate differences in respiratory efficiency between *M. lauensis* COX and other species.

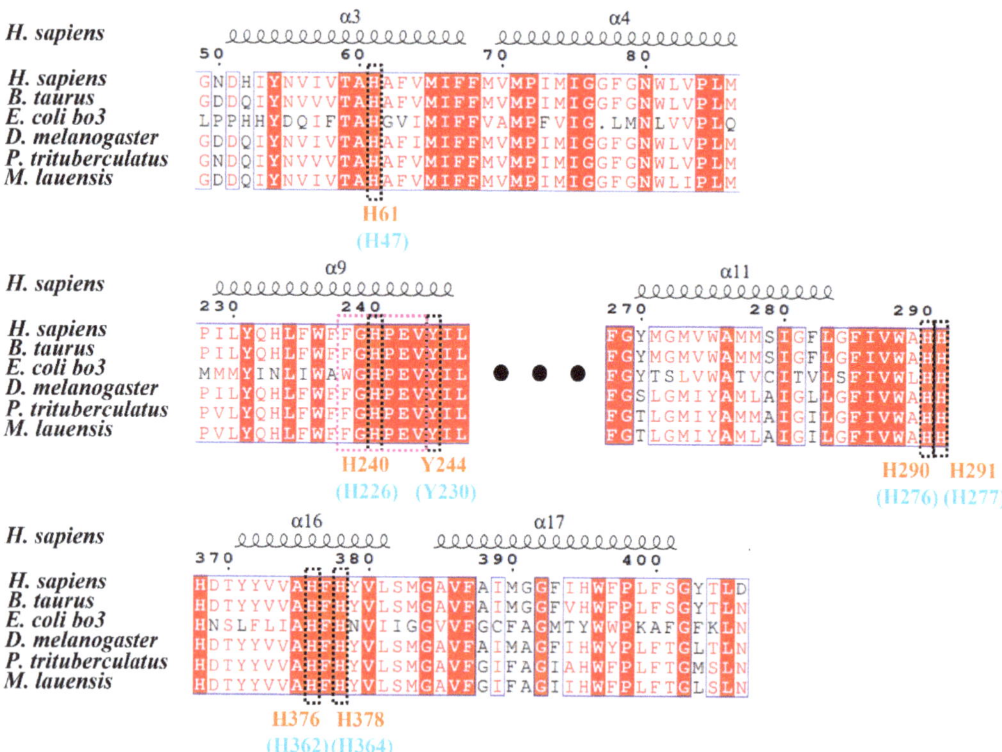

Figure 4. Sequence alignment of the COX1 type A1 motif in six species (accession numbers in parentheses). *H.*, *Homo* (AGW78696), *B.*, *Bos* (P00396), *E.*, *Escherichia* (P0ABI8), *D.*, *Drosophila* (AAF77227), *P.*, *Portunus* (QPD06751), *M.*, *Munidopsis* (QFG40073). Red boxes and white characters indicate strict sequence identities, and red characters show residues with high similarity. Black-dashed boxes and orange labels indicate the heme-interacting residues of human COX1 (H61, H240, Y244, H290, H291, H376, and H378). Cyan labels represent the H47, H226, Y230, H276, H277, H362, and H364 residues of *M. lauensis* COX1. Magenta-dashed boxes represent the COX1 type A1 motif (-XGHPEV-). Secondary structures are represented above the alignment.

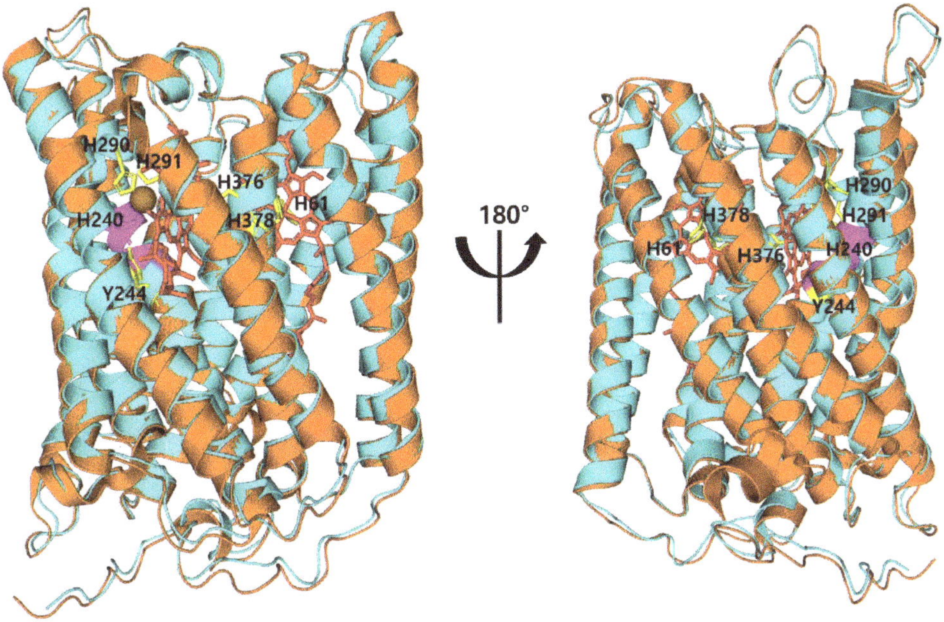

Figure 5. Comparison of predicted structure models for *M. lauensis* COX1 and human COX1 (PDB: 5z62). Human and *M. lauensis* COX1 are colored in orange and cyan, respectively. Heme-binding amino acids (yellow), heme (red), Cu_B (brown), and the type A1 motif (magenta) are also shown. The type A1 motif of helix 9 comprises -XGHPEV- amino acid sequences in all species. The heme-binding regions of *M. lauensis* and five species have the same location as the human protein. Indeed, amino acid sequences around the heme-binding regions of all species are highly homogeneous. The type A1 motif is strictly conserved between human and *M. lauensis*.

4. Conclusions

Mitochondria are predominantly responsible for aerobic metabolism, and functional restrictions relate to adaptive evolutionary mechanisms. Our findings indicate that our isolates of the decapod crustacean *M. lauensis* from a deep-sea hydrothermal vent in the Indian Ocean are new. Previous records were from vents of the southwest Pacific. In addition, the arrangement of heme-binding regions and type A1 motifs of *M. lauensis* are identical to those of six other species. Dispersal of the same species to geologically separated hydrothermal vents with conserved heme-binding regions in mitochondria suggest that hydrothermal species might have evolved from shallow sea environments and a dispersal corridor. Our study provides useful information for additional studies on hydrothermal vents. Furthermore, our study guides future work to characterize the dispersal corridor between the western Pacific and Indian Oceans.

Author Contributions: Conceptualization, H.-s.H., B.P. and T.K.; methodology, H.-s.H., B.C. and J.C.; investigation, B.C. and J.C.; data curation, H.-s.H., B.C. and J.C.; writing—original draft preparation, H.-s.H., B.C., J.C., B.P. and T.K.; writing—review and editing, B.P. and T.K.; visualization, H.-s.H., J.C., B.P. and T.K.; project administration, B.C. and T.K.; funding acquisition, T.K. All authors have read and agreed to the published version of the manuscript.

Funding: This research was a part of the project titled "Understanding the deep-sea biosphere on seafloor hydrothermal vents in the Indian Ridge" (No. 20170411) funded by the Ministry of Oceans and Fisheries, Korea. It was also partly funded by the Korea Institute of Ocean & Technology (KIOST) research program (Grant no. PE99985). B. Park was supported by Eulji University in 2020.

Institutional Review Board Statement: Not applicable.

Informed Consent Statement: Not applicable.

Data Availability Statement: All data used are presented in the text, figures, and tables.

Acknowledgments: We are grateful to Ok Hwan Yu and Dong Sung Kim for helping collection of samples.

Conflicts of Interest: The authors declare no conflict of interest.

References

1. Little, C.T.S.; Vrijenhoek, R.C. Are hydrothermal vent animals living fossils? *Trends Ecol. Evol.* **2003**, *18*, 582–588. [CrossRef]
2. Mullineaux, L.S.; Metaxas, A.; Beaulieu, S.E.; Bright, M.; Gollner, S.; Grupe, B.M.; Herrera, S.; Kellner, J.B.; Levin, L.A.; Mitarai, S.; et al. Exploring the ecology of deep-sea hydrothermal vents in a metacommunity framework. *Front. Mar. Sci.* **2018**, *5*, 49. [CrossRef]
3. McArthur, A.G.; Tunnicliffe, V. Relics and Antiquity Revisited in The Modern Vent Fauna. In *Modern Ocean Floor Processes and the Geological Record*; Mills, R., Harrison, K., Eds.; Special Publication of the Geological Society of London: Burlington, UK, 1998; pp. 271–291.
4. Van Dover, C.L. *The Ecology of Deep-Sea Hydrothermal Vents*; Princeton University Press: Oxford, UK, 2000.
5. Van Dover, C.L. Ecology of Mid Atlantic Ridge Hydrothermal Vents. In *Hydrothermal Vents and Process*; Parson, C.M., Dixon, D.R., Eds.; Geological Society of London, Special Publication 887: London, UK, 1995; pp. 257–294.
6. Gebruk, A.V.; Galkin, S.V.; Vereshchaka, A.L.; Moskalev, L.I.; Southward, A.J. Ecology and biogeography of the hydrothermal vent fauna of the Mid-Atlantic Ridge. *Adv. Mar. Biol.* **1997**, *32*, 93–144.
7. Tunnicliffe, V. The biology of hydrothermal vents: Ecology and evolution. *Oceanogr. Mar. Biol. Annu. Rev.* **1991**, *29*, 319–407.
8. Hashimoto, J.; Ohta, S.; Fujikura, K.; Miura, T. Microdistribution pattern and biogeography of the hydrothermal vent communities of the Minami-Ensie Knoll in the mid-Okinawa trough, Western Pacific. *Deep Sea Res. I* **1995**, *42*, 577–598. [CrossRef]
9. Hashimoto, J.; Ohta, S.; Gamo, T.; Chiba, H.; Yamaguchi, T.; Tsuchida, S.; Okudaira, T.; Watabe, H.; Yamanaka, T.; Kitazawa, M. First hydrothermal vent communities from the Indian Ocean discovered. *Zool. Sci.* **2001**, *18*, 717–721. [CrossRef]
10. Van Dover, C.L.; Humphris, S.E.; Fornari, D.; Cavanaugh, C.M.; Collier, R.; Goffredi, S.K.; Hashimoto, J.; Lilley, M.D.; Reysenbach, A.L.; Shank, T.M.; et al. Biogeography and ecological setting of Indian Ocean hydrothermal vents. *Science* **2001**, *294*, 818–823. [CrossRef] [PubMed]
11. Martin, J.W.; Haney, T.A. Decapod crustaceans from hydrothermal vents and cold seeps: A review through 2005. *Zool. J. Linn. Soc.* **2005**, *145*, 445–522. [CrossRef]
12. Yang, C.H.; Tsuchida, S.; Fujikura, K.; Fujiwara, Y.; Kawato, M.; Chan, T.Y. Connectivity of the squat lobsters *Shinkaia crosnieri* (Crustacea: Decapoda: Galatheidae) between cold seep and hydrothermal vent habitats. *Bull. Mar. Sci.* **2016**, *92*, 17–31. [CrossRef]
13. Hui, M.; Song, C.; Liu, Y.; Li, C.; Cui, Z. Exploring the molecular basis of adaptive evolution in hydrothermal vent crab *Austinograea alayseae* by transcriptome analysis. *PLoS ONE* **2017**, *12*, e0178417. [CrossRef] [PubMed]
14. Sun, S.E.; Sha, Z.; Wang, Y. Divergence history and hydrothermal vent adaptation of decapod crustaceans: A mitogenomic perspective. *PLoS ONE* **2019**, *14*, e0224373.
15. Baba, K. Deep-sea chirostylid and galatheid Crustaceans (Decapoda: Anomura) from the Indo-Pacific, with a list of species. *Galathea Rep.* **2005**, *20*, 317.
16. Macpherson, E. Species of the genus *Munidopsis* Whiteaves, 1784 from the Indian and Pacific oceans and reestablishment of the genus *Galacantha* A. Milne-Edwards, 1880 (Crustacea, Decapoda, Galatheidae). *Zootaxa* **2007**, *1417*, 1–135. [CrossRef]
17. Williams, A.B. New Marine decapod crustaceans from waters influenced by hydrothermal discharge, brine and hydrocarbon seepage. *Fish. Bull.* **1988**, *86*, 213–287.
18. Williams, A.B.; Baba, K. New squat lobsters (Galatheidae) from the Pacific Ocean: Mariana Back Arc, East Pacific Rise and Cascadian Basin. *Fish. Bull.* **1989**, *87*, 899–910.
19. Baba, K. A new squat lobster (Decapoda: Anomura: Galatheidae) from an active thermal vent area in the North Fiji Basin, SW Pacific. *Crust. Res.* **1995**, *24*, 188–193. [CrossRef]
20. Macpherson, E.; Segonzac, M. Species of genus *Munidopsis* (Decapoda, Anomura, Galatheidae) from the deep Atlantic Ocean, including cold seeps and hydrothermal vent area. *Zootaxa* **2005**, *1095*, 1–60. [CrossRef]
21. Cubelio, S.S.; Tsuchida, S.; Hendrickx, M.E.; Kado, R.; Watanabe, S. A new species of vent associated *Munidopsis* (Crustacea: Decapoda: Anomura: Galatheidae) from the Western Pacific, with notes on its genetic identification. *Zootaxa* **2007**, *1435*, 25–36. [CrossRef]
22. Cubelio, S.S.; Tsuchida, S.; Watanabe, S. New Species of *Munidopsis* (Decapoda: Anomura: Galatheidae from hydrothermal vent areas of Indian and Pacific Oceans. *J. Mar. Biol. Assoc. UK* **2008**, *88*, 111–117. [CrossRef]
23. Cubelio, S.S.; Tsuchida, S.; Watanabe, S. Vent Associated *Munidopsis* (Decapoda: Anomura: Galatheidae) from Brothers Seamount, Kermadec Arc, Southwest Pacific, with Description of One New Species. *J. Crustac. Biol.* **2007**, *3*, 513–519. [CrossRef]

24. Thaler, A.D.; Plouviez, S.; Saleu, W.; Alei, F.; Jacobson, A.; Boyle, E.A.; Van Dover, C.L. Comparative population structure of two deep-sea hydrothermal-vent-associated decapods (*Chorocaris* sp. 2 and *Munidopsis lauensis*) from southwestern Pacific back-arc basins. *PLoS ONE* **2014**, *97*, e101345. [CrossRef] [PubMed]
25. Baba, K.; Macpherson, E.; Poore, G.C.; Ahyong, S.T.; Bermudez, A.; Cabezas, P.; Schnabel, K.E. Catalogue of squat lobsters of the world (Crustacea: Decapoda: Anomura families Chirostylidae, Galatheidae and Kiwaidae). *Zootaxa* **2008**, *1905*, 1–220. [CrossRef]
26. Lin, C.W.; Tsuchida, S.; Lin, S.; Berndt, C.; Chan, T.Y. *Munidopsis lauensis* Baba & de Saint Laurent, 1992 (Decapoda, Anomura, Munidopsidae), a newly recorded squat lobster from a cold seep in Taiwan. *Zootaxa* **2013**, 92–96. [CrossRef]
27. Goffredi, S.K.; Warén, A.; Orphan, V.J.; Van Dover, C.L.; Vrijenhoek, R.C. Novel forms of structural integration between microbes and a hydrothermal vent gastropod from the Indian Ocean. *Appl. Environ. Microbiol.* **2004**, *70*, 3082–3090. [CrossRef] [PubMed]
28. Shen, Y.Y.; Shi, P.; Sun, Y.B.; Zhang, Y.P. Relaxation of selective constraint on avian mitochondrial DNA following the degeneration of flight ability. *Genome Res.* **2009**, *19*, 1760–1765. [CrossRef] [PubMed]
29. Shen, Y.Y.; Liang, L.; Zhu, Z.H.; Zhou, W.P.; Irwin, D.M.; Zhang, Y.P. Adaptive evolution of energy metabolism genes and the origin of flight in bats. *Proc. Natl. Acad. Sci. USA* **2010**, *107*, 8666–8671. [CrossRef] [PubMed]
30. Luo, Y.; Gao, W.; Gao, Y.; Tang, S.; Huang, Q.; Tan, X.; Chen, J.; Huang, T. Mitochondrial genome analysis of *Ochotona curzoniae* and implication of cytochrome c oxidase in hypoxic adaptation. *Mitochondrion* **2008**, *8*, 352–357. [CrossRef]
31. Gering, E.J.; Opazo, J.C.; Storz, J.F. Molecular evolution of cytochrome b in high- and low-altitude deer mice (genus Peromyscus). *Heredity* **2009**, *102*, 226–235. [CrossRef]
32. Scott, G.R.; Schulte, P.M.; Egginton, S.; Scott, A.L.; Richards, J.G.; Milsom, W.K. Molecular evolution of cytochrome c oxidase underlies high-altitude adaptation in the bar-headed goose. *Mol. Biol. Evol.* **2011**, *28*, 351–363. [CrossRef]
33. Wang, Z.; Yonezawa, T.; Liu, B.; Ma, T.; Shen, X.; Su, J.; Guo, S.; Hasegawa, M.; Liu, J. Domestication relaxed selective constraints on the yak mitochondrial genome. *Mol. Biol. Evol.* **2011**, *28*, 1553–1556. [CrossRef]
34. Gu, M.; Dong, X.; Shi, L.; Shi, L.; Lin, K.; Huang, X.; Chu, J. Differences in mtDNA whole sequence between Tibetan and Han populations suggesting adaptive selection to high altitude. *Gene* **2012**, *496*, 37–44. [CrossRef] [PubMed]
35. Zhou, T.; Shen, X.; Irwin, D.M.; Shen, Y.; Zhang, Y. Mitogenomic analyses propose positive selection in mitochondrial genes for high-altitude adaptation in galliform birds. *Mitochondrion* **2014**, *18*, 70–75. [CrossRef] [PubMed]
36. da Fonseca, R.R.; Johnson, W.E.; O'Brien, S.J.; Ramos, M.J.; Antunes, A. The adaptive evolution of the mammalian mitochondrial genome. *BMC Genomics* **2008**, *9*, 119. [CrossRef] [PubMed]
37. Siebenaller, J.F.; Garrett, D.J. The effects of the deep-sea environment on transmembrane signaling. *Comp. Biochem. Physiol. B Biochem. Mol. Biol.* **2002**, *131*, 675–694. [CrossRef]
38. Hebert, P.D.; Gregory, T.R. The promise of DNA barcoding for taxonomy. *Syst. Biol.* **2005**, *54*, 852–859. [CrossRef] [PubMed]
39. Pereira, M.M.; Santana, M.; Teixeira, M. A novel scenario for the evolution of heme–copper oxygen reductases. *Biochim. Biophys. Acta Bioenerg.* **2001**, *1505*, 185–208. [CrossRef]
40. Ahyong, S.T.; Andreakis, N.; Taylor, J. Mitochondrial phylogeny of the deep-sea squat lobsters, Munidopsidae (Galatheoidea). *Zool. Anz.* **2011**, *250*, 367–377. [CrossRef]
41. Folmer, O.; Black, M.; Hoeh, W.; Lutz, R.; Vrijenhoek, R. DNA primers for amplification of mitochondrial cytochrome c oxidase subunit I from diverse metazoan invertebrates. *Mol. Mar. Biol. Biotechnol.* **1994**, *3*, 294–299. [PubMed]
42. Katoh, K.; Misawa, K.; Kuma, K.I.; Miyata, T. MAFFT: A novel method for rapid multiple sequence alignment based on fast Fourier transform. *Nucleic Acid. Res.* **2002**, *30*, 3059–3066. [CrossRef] [PubMed]
43. Kimura, M. A Simple method for estimating evolutionary rates of base substitutions through comparative studies of nucleotide sequences. *J. Mol. Evol.* **1980**, *16*, 111–120. [CrossRef] [PubMed]
44. Kumar, S.; Stecher, G.; Li, M.; Knyaz, C.; Tamura, K. MEGA X: Molecular evolutionary genetics analysis across computing platforms. *Mol. Biol. Evol.* **2018**, *35*, 1547–1549. [CrossRef] [PubMed]
45. Lanfear, R.; Calcott, B.; Ho, S.Y.W.; Guindon, S. PartitionFinder: Combined selection of partitioning schemes and substitution models for phylogenetic analyses. *Mol. Biol. Evol.* **2012**, *29*, 1695–1701. [CrossRef] [PubMed]
46. Bouckaert, R.; Heled, J.; Kühnert, D.; Vaughan, T.; Wu, C.H.; Xie, D.; Suchard, M.A.; Rambaut, A.; Drummond, A.J. BEAST 2: A software platform for Bayesian evolutionary analysis. *PLoS Comput. Biol.* **2014**, *10*, e1003537. [CrossRef] [PubMed]
47. Drummond, A.J.; Ho, S.Y.W.; Phillips, M.J.; Rambaut, A. Relaxed phylogenetics and dating with confidence. *PLoS Biol.* **2006**, *4*, e88. [CrossRef] [PubMed]
48. Rambaut, A.; Drummond, A.J.; Xie, D.; Baele, G.; Suchard, M.A. Posterior summarization in Bayesian phylogenetics using Tracer 1.7. *Syst. Biol.* **2018**, *67*, 901. [CrossRef] [PubMed]
49. Helfrich, P.; Rieb, E.; Abrami, G.; Lücking, A.; Mehler, A. TreeAnnotator: Versatile Visual Annotation of Hierarchical Text Relations. In Proceedings of the Eleventh International Conference on Language Resources and Evaluation (LREC 2018, European Language Resources Association (ELRA)), Paris, France, 7–12 May 2018.
50. Sievers, F.; Higgins, D.G. Clustal omega. *Curr. Protoc. Bioinform.* **2014**, *48*, 3–13. [CrossRef] [PubMed]
51. Gouet, P.; Robert, X.; Courcelle, E. ESPript/ENDscript: Extracting and rendering sequence and 3D information from atomic structures of proteins. *Nucleic Acids Res.* **2003**, *31*, 3320–3323. [CrossRef] [PubMed]
52. Server, S.M. SWISS-MODEL SERVER (https://swissmodel.expasy.org). *Int. J. Mol. Sci.* **2014**, *15*, S10.
53. Schrödinger, L.; DeLano, W. PyMOL. 2020. Available online: http://www.pymol.org/pymol (accessed on 3 January 2022).

54. German, C.R.; Baker, E.T.; Mevel, C.; Tamaki, K. Hydrothermal activity along the southwest Indian ridge. *Nature* **1998**, *395*, 490–493. [CrossRef]
55. Kim, M.; Kang, J.-H.; Kim, D. Holoplanktonic and meroplanktonic larvae in the surface waters of the Onnuri Vent Field in the Central Indian Ridge. *J. Mar. Sci. Eng.* **2022**, *10*, 158. [CrossRef]
56. Wikström, M.; Krab, K.; Sharma, V. Oxygen activation and energy conservation by cytochrome c oxidase. *Chem. Rev.* **2018**, *118*, 2469–2490. [CrossRef] [PubMed]

Journal of
Marine Science and Engineering

MDPI

Article

Trimethylamine N-Oxide (TMAO) and Trimethylamine (TMA) Determinations of Two Hadal Amphipods

Qi Liu [1,2], Shouwen Jiang [3], Wenhao Li [3], Binbin Pan [1] and Qianghua Xu [1,2,4,*]

1 Shanghai Engineering Research Center of Hadal Science & Technology, College of Marine Sciences, Shanghai Ocean University, Shanghai 201306, China; m190200531@st.shou.edu.cn (Q.L.); bbpan@shou.edu.cn (B.P.)
2 Key Laboratory of Sustainable Exploitation of Oceanic Fisheries Resources, Ministry of Education, College of Marine Sciences, Shanghai Ocean University, 999 Huchenghuan Road, Lingang New City, Shanghai 201306, China
3 Key Laboratory of Aquaculture Resources and Utilization, Ministry of Education, College of Fisheries and Life Sciences, Shanghai Ocean University, Shanghai 201306, China; swjiang@shou.edu.cn (S.J.); wh-li@shou.edu.cn (W.L.)
4 National Distant-Water Fisheries Engineering Research Center, Shanghai Ocean University, Shanghai 201306, China
* Correspondence: qhxu@shou.edu.cn

Abstract: Hadal trenches are a unique habitat with high hydrostatic pressure, low temperature and scarce food supplies. Amphipods are the dominant scavenging metazoan species in this ecosystem. Trimethylamine (TMA) and trimethylamine oxide (TMAO) have been shown to play important roles in regulating osmotic pressure in mammals, hadal dwellers and even microbes. However, the distributions of TMAO and TMA concentrations of hadal animals among different tissues have not been reported so far. Here, the TMAO and TMA contents of eight tissues of two hadal amphipods, *Hirondellea gigas* and *Alicella gigantea* from the Mariana Trench and the New Britain Trench, were detected by using the ultrahigh performance liquid chromatography–mass spectrometry (UPLC-MS/MS) method. Compared with the shallow water Decapoda, *Penaeus vannamei*, the hadal amphipods possessed significantly higher TMAO concentrations and a similar level of TMA in all the detected tissues. A higher level of TMAO was detected in the external organs (such as the eye and exoskeleton) for both of the two hadal amphipods, which indicated that the TMAO concentration was not evenly distributed, although the same hydrostatic pressure existed in the outer and internal organs. Moreover, a strong positive correlation was found between the concentrations of TMAO and TMA in the two hadal amphipods. In addition, evolutionary analysis regarding FMO3, the enzyme to convert TMA into TMAO, was also conducted. Three positive selected sites in the conserved region and two specific mutation sites in two conserved motifs were found in the *A. gigantea* FMO3 gene. Combined together, this study supports the important role of TMAO for the environmental adaptability of hadal amphipods and speculates on the molecular evolution and protein structure of FMO3 in hadal species.

Keywords: hadal; amphipod; TMAO; FMO3; molecular evolution

1. Introduction

The hadal zone is the deepest area of the ocean, extending from 6000 to 11,000 m in depth from the ocean surface and accounts for 45 percent of the total ocean depth range [1]. The hadal region represents 1–2% of the global benthic area, consisting mainly of trenches characterized by high hydrostatic pressure (HHP), low temperature, darkness, and low organic matter [2–6]. As the deepest ecosystem, the hadal zone is one of the most unique ecosystems on earth. Compared with the shallow sea, HHP is a major feature that distinguishes hadal trenches from other ecosystems. HHP is unique in that it manifests as the largest continuous, stable gradient of any stressor in the hadal zone, reaching about 100 Mpa in the deepest area [7,8].

However, despite such harsh conditions, there are still many species living in hadal habitats. According to extensive hadal sampling records, amphipods are the most dominant decomposers in hadal environments, which are widely distributed in many trenches, with habitats up to 10,000 m deep, and are easy to trap [9–11]. Among them, *Alicella gigantea* and *Hirondellea gigas* are reported as the most common amphipods in the hadal environment. *A. gigantea* is the largest known amphipod, reaching 340 mm in length [12,13]; *H. gigas* can even be found and captured at depths of 11,000 m [14]. Therefore, studies regarding hadal amphipods' adaptations can give us insights into the adaptive mechanisms of other species in hadal environments.

Generally, HHP may break protein structure [15], cause DNA breakage and damage [16] and reduce cell membrane fluidity [17]. To survive in the hadal environment, the extrinsic adaptation mechanism and intrinsic adaptation were needed [18]. The extrinsic adaptation includes different molecular and chemical chaperones. The molecular chaperones such as the heat-shock proteins (HSPs) are well known for resisting apoptosis, assisting transmembrane transport, and helping protein folding and transporting [19,20]. For deep-sea species, chemical chaperones, especially the organic osmolytes, have been reported to play important roles in adapting to HHP. These are the most important small molecules found in many organisms to maintain cell volume and protein function when facing stress, mainly including ammonia nitrogen compounds, amino acid derivatives, polyols, sugars and urea [21]. TMA (trimethylamine) is an important ammonia–nitrogen compound which is mainly formed by the consumption of carnitine choline by gut microorganisms [22]. TMAO (trimethylamine oxide) is an important osmotic regulator produced by TMA oxidation under the action of the flavin monooxygenases (FMOs) [23]. TMAO is a powerful protein stabilizer commonly found in marine fish muscle tissue, which can mitigate the effects of hydrostatic pressure on protein stability and restore denatured proteins to their natural structure [24]. TMAO is a universal protein stabilizer and counteractant for resisting urea damage [21]. TMAO can also act as a piezolyte, which could bind with water, prevent water molecules clumping together and avoid the protein aggregations [25,26].

TMAO concentrations are reported to increase gradually with the increase in depth, suggesting its important function for the hadal species [21,27]. In chondrichthyan species (Chondrichthyes), TMAO content increases from ~150 mmol/kg to ~200 mmol/kg with the depth increasing from 500 to 1500 m [28]. In teleost fish, muscle TMAO content ranges from less than 50 mmol/kg in shallow species to nearly 400 mmol/kg when the depth is nearly 8000 m [24]. As for invertebrates (such as squid, decapods and amphipods), TMAO has been found to increase linearly with depth. For example, TMAO content in muscle of the decapod *Pandalus danae* increases linearly from 76 mmol/kg to 299 mmol/kg with the depth increasing from 0 to 2850 m [29]. As for amphipods, TMAO concentrations were reported ranging from less than 15 mmol/kg in shallow sea species to above 80 mmol/kg in a deep-sea sample [30,31]. It should be noticed that, for a decapod at 6000 m depth, the TMAO content in white muscle is about 270 mmol/kg, whereas the TMAO in hemolymph is only 12.9 mmol/kg [32], which indicates the great variations of TMAO concentration between tissues. However, the distributions of TMAO concentrations of hadal species among different tissues have not been reported so far.

Hadal environment intrinsic adaptation mainly refers to the evolutionary adaptation of the amino acid sequence of protein itself, which includes amino acid substitutions in some specific sites and protein structure changes [33,34]. FMOs belong to a subfamily of B monooxygenases and are conserved in all phyla. Their main function is to add molecular oxygen to lipophilic compounds [35,36]. There are five different functional FMO in adult mammals, numbered from one to five, and FMO3 is the most important TMA oxidase expressed in the liver [34,37]. The oxidation activity of FMO is very efficient because it does not require the presence of substrates to initiate the catalytic cycle. The prosthetic group was offered by flavin adenine dinucleotide (FAD), nicotinamide adenine dinucleotide phosphate hydrogen (NADPH) was used as a hydride donor, and oxygen existed as cosubstrate [38]. The FMO expression could help cells to be resistant to multiple stressors, including high-pressure stress, heavy metal contaminations, free radical generator paraquat, UV radiation and the mitochondrial toxin rotenone [34,37]. However, little is known about the FMO in hadal amphipods.

Therefore, we decided to measure the TMAO and TMA content in two hadal amphipod species, *A. gigantea* and *H. gigas*, and compared them with the content in a shallow-water decapod species, *Penaeus vannamei*. Evolutionary analysis regarding to the hadal amphipods' FMO3 was also conducted. Our research on TMAO and TMA content in different tissues in the hadal amphipods provides new insights into the possible molecular adaptation mechanisms of the hadal amphipods.

2. Materials and Methods

2.1. Source of the Sample

Two hadal amphipod species, *Hirondellea gigas* and *Alicella gigantea*, were collected from the Mariana Trench (10,839 m, 11.38° N 142.42° E) and the New Britain Trench (8824 m, 7.02° S 149.16° E) in the west Pacific Ocean. Amphipod samples were collected by the autonomous deep-ocean lander vehicle launched from the "Zhang Jian" research vessel over the course of four sampling campaigns [38]. The detailed information about the lander vehicle and sampling were described in our previous study [38]. Once collected on-board, each amphipod sample was placed in a separate zip-lock bag and was immediately frozen upon recovery at −80 °C. One shallow-water decapod species, *P. vannamei*, collected from Qingdao coastal area with the depth ranging from 50~75 m, was also used in this study.

2.2. Pretreatment of Experimental Samples

Hirondella gigas, *Alicella gigantea* and *Penaeus vannamei* were dissected and eight tissues (eye, brain, exoskeleton, gonad, fat, gut, muscle and liver) were obtained for the following experiment. Each tissue was extracted from 10~30 individuals and the measurements for each tissue were repeated 9 times.

Twenty milligrams of samples from each tissue were stored in a centrifugal tube, and 1200 μL solution (CH3OH: water = 4:1) was added. Then, vibration crushing was performed on a high-throughput tissue crusher to crush the tissue samples. Ice bath ultrasonic extractions were subsequently conducted three times. The tissue suspensions were stored in the refrigerator at −20 °C for 20 min and centrifuged for 10 min at 13,000 rpms at 4 °C. The supernatant was collected and diluted 20 times. Four hundred microliters of diluted liquid were sucked out and then incorporated into the sample by chromatography using the LC-MS and stored in a 4 °C refrigerator for further LC-MS, UPLC tests and analysis.

2.3. Settings of the UPLC-MS/MS Parameters

The quantifications of TMAO and TMA were performed by UPLC-MS/MS (WATERS Inc., Milford, MA, USA). The chromatographic separation was carried out on an Infinity II HILIC column. The flow rate was maintained at 0.2 mL/min, and the column was heated to 30 °C. The instrument parameters for WATERS UPLC-MS/MS analysis are as follows: nitrogen drying gas temperature 300 °C, nitrogen sheath gas temperature 250 °C, nitrogen drying gas flow 5 L min^{-1}, nitrogen sheath gas flow 11 L min^{-1}, capillary voltage 3500 V, nebulizer pressure 45 psi and nozzle voltage 500 V. The information regarding untested compounds and internal standards was detected by characteristic precursor-product ion transitions.

The detection was carried out by using a triple quadrupole mass spectrometer in the positive ion mode in the multiple reaction monitoring (MRM) modes. The sample concentrations were determined from calibration curves using a peak area ratio of the analyte to its isotope [39].

2.4. Evolutionary Analysis and Protein Crystal Structure Prediction of FMO3

The BLAST program of National Center for Biotechnology Information (NCBI) was used to search deduced amino acid sequences of FMO3 derived from selected crustacean.

ClustalW in MEGA X was used for multiple sequence alignment. Phylogenetic tree was constructed by ML method with the bootstrap of 1000 and Figtree was used for beautification. PFAM, Interpro and SMART databases were used to annotate the deduced FMO3 of *A. gigantea* and *P. vannamei* for structural domain analysis. Conserved motif sequences were predicted from PFAM database. Evolutionary analysis was conducted through the Branch site model in Preset Mode in EasyCodeml (Version 1.4) [40]. Through the likelihood ratio test (LRT) test, the crystal structure model of FMO3 and FAD ligand was carried out by the I-TASSER database.

3. Results

3.1. TMAO and TMA Concentrations of the Two Hadal Amphipods

In this study, two compounds (TMAO and TMA) in eight tissues (eye, brain, muscle, exoskeleton, gonad, fat, gut, muscle and liver) of two hadal amphipod species (*A. gigantea* and *H. gigas*) and one shallow-water decapod, *P. vannamei*, were determined. All the TMAO and TMA concentrations are shown in Tables 1 and 2, respectively. It was shown that the order of TMAO mean content level is *A. gigantea* > *H. gigas* > *P. vannamei* (Figure 1). The two hadal amphipods (*A. gigantea* and *H. gigas*) have a higher level of TMAO, even up to 10 times higher than that of their shallow-water counterpart (*P. vannamei*) (Figure 1a). However, the order of TMA mean level of the species is *P. vannamei* > *H. gigas* > *A. gigantea*, which was quite different from the TMAO (Figure 1b).

Table 1. The trimethylamine oxide (TMAO) content (unit: mmol/kg wet weight; mean ± SD) across eight tissues in three species.

Tissue	Tissue Trimethylamine Oxide (TMAO) Content (mmol/kg)			Significant Difference		
	A. gigantea	*H. gigas*	*P. vannamei*	Ag-Hg	Ag-Pv	Hg-Pv
eye	68.52 ± 12.36	42.80 ± 2.26	2.49 ± 0.44	***	***	***
brain	66.69 ± 14.87	18.99 ± 7.93	1.96 ± 0.75	***	***	**
exoskeleton	55.47 ± 11.76	31.62 ± 9.56	2.75 ± 0.80	***	***	***
gonad	43.87 ± 1.16	15.70 ± 5.81	1.82 ± 0.28	***	***	***
fat	40.55 ± 7.12	22.22 ± 5.13	2.88 ± 0.13	***	***	***
gut	34.53 ± 19.78	28.12 ± 8.38	2.22 ± 0.32	-	***	***
muscle	27.36 ± 16.02	23.09 ± 21.13	2.48 ± 0.65	-	**	*
liver	20.48 ± 12.20	16.24 ± 6.44	2.84 ± 0.32	**	****	**

* represents significant difference ($p < 0.05$), ** represents significant difference ($p < 0.01$), and *** represents significant difference ($p < 0.001$).

Table 2. The trimethylamine (TMA) content (unit: mmol/kg wet weight; mean ± SD) across eight tissues in three species.

Tissue	Tissue Trimethylamine (TMA) Content (mmol/kg)			Significant Difference		
	A. gigantea	*H. gigas*	*P. vannamei*	Ag-Hg	Ag-Pv	Hg-Pv
eye	5.48 ± 3.68	7.99 ± 0.40	7.89 ± 1.55	-	-	-
brain	4.02 ± 1.60	3.40 ± 0.85	6.52 ± 0.99	-	***	***
exoskeleton	6.67 ± 3.90	4.90 ± 2.51	4.97 ± 2.09	-	-	-
gonad	5.65 ± 4.52	2.98 ± 0.66	7.77 ± 2.98	-	-	**
fat	3.13 ± 0.93	3.08 ± 0.91	3.67 ± 2.50	-	-	-
gut	3.55 ± 2.37	3.98 ± 1.35	12.93 ± 5.26	-	***	***
muscle	2.09 ± 1.11	7.12 ± 12.41	3.65 ± 0.96	-	**	**
liver	3.08 ± 1.13	3.07 ± 0.82	7.08 ± 1.82	-	***	***

** represents significant difference ($p < 0.01$), and *** represents significant difference ($p < 0.001$).

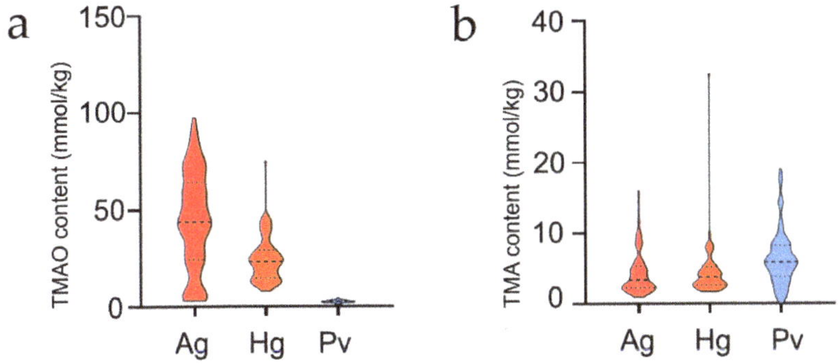

Figure 1. (**a**,**b**) Violin diagram of total TMAO and TMA content of two hadal amphipods and a shallow water decapods *P. vannamei*. Red represents *A. gigantea*, orange represents *H. gigas*, and blue represents *P. vannamei*.

Except the comparison with *P. vannamei*, we also compared our study with the reported TMAO concentrations in amphipod species [30,41]. The comparisons are shown in Table 3. Generally, TMAO concentrations in the hadal *A. gigantea* and *H. gigas* are higher than those of the shallow amphipod in Lake Baikal (Table 3). Meanwhile, it can be easily concluded that the TMAO concentration increases generally with increasing depth. As for hadal *H. gigas*, the TMAO concentration of the combined tissues (our study) is closed to the data of the whole body in Downing's study [30].

Table 3. The TMAO concentrations of amphipods from fresh water and marine environments (unit: mmol/kg wet weight).

Species	Tissue	Depth	Location	TMAO	Reference
Megalorchestia columbiana	whole body	−1 m	Sandy beach	15	Downing et al., 2018
Anisogammarus pugettensi	whole body	0.1 m	Sandy beach	12	Downing et al., 2018
Acanthogammarus lappadeus	muscle	50 m	Lake Baikal	6.0	Zerbst et al., 2005
Acanthogammarus grewingki	muscle	170–930 m	Lake Baikal	18.1–28.4	Zerbst et al., 2005
Acanthogammarus albus	muscle	200 m	Lake Baikal	16.1	Zerbst et al., 2005
Scypholanceola aestiva	whole body	763 m	Tidepool	17	Downing et al., 2018
Acanthogammarus reicherti	muscle	930 m	Lake Baikal	31.6	Zerbst et al., 2005
Ceratogammarus dybowskii	muscle	930–1170 m	Lake Baikal	43.3–47.3	Zerbst et al., 2005
Parapallasea lagowskii	muscle	1170 m	Lake Baikal	32.0	Zerbst et al., 2005
Valettietta sp	whole body	1561 m	Southwest of Oahu	30	Downing et al., 2018
Paralicella tenupies	whole body	3569–4779 m	Kermadec Trench	45–50	Downing et al., 2018
Eurythenes gryllus	whole body	3865–4817 m	Kermadec Trench	26–42	Downing et al., 2018
Cyclocaris	whole body	4897 m	Marina Trench	25	Downing et al., 2018
Bathycallisoma schellenbergi	whole body	5958–9198 m	Kermadec Trench	46–82	Downing et al., 2018
Abyssorchomene musculosus	whole body	6081 m	Marina Trench	30	Downing et al., 2018
Hirondellea gigas	whole body	6974–10,991 m	Marina Trench	38–64	Downing et al., 2018
Hirondellea dubia	whole body	7515–10,005 m	Kermadec Trench	56–75	Downing et al., 2018
Princaxelia jamiesoni	whole body	8189 m	Marina Trench	43	Downing et al., 2018
Alicella gigantea	combined	8824 m	New Britain Trench	20–68	This study
Hirondellea gigas	combined	10,839 m	Marina Trench	15.7–42.8	This study

Note: Downing et al., 2018 [32]; Zerbst et al., 2005 [41].

The mean TMAO and TMA content (mmol/kg) and standard deviations of eight tissues in three species are shown in Figure 2. It was clearly shown that the hadal group (*H. gigas* and *A. gigantea*) possessed significantly higher levels of TMAO than the shallow-water *P. vannamei* across all the eight tissues (Figure 2a). Different from TMAO, only five tissues (brain, gonad, gut, liver, muscle) exhibited significant differences between the two groups (Figure 2b).

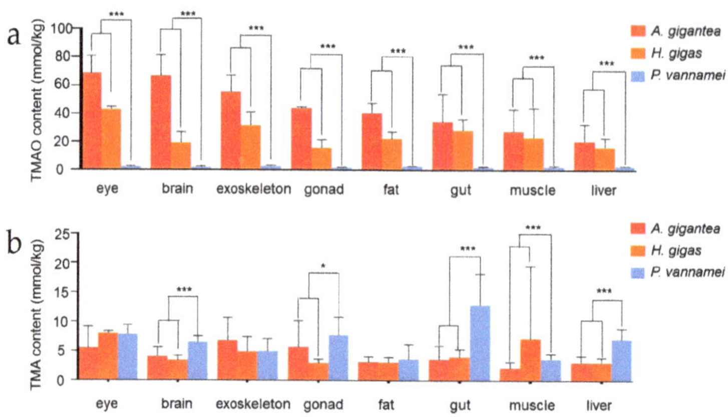

Figure 2. (**a**,**b**) TMAO and TMA content in eight tissues (eye, brain, muscle, exoskeleton, gonad, fat, gut, muscle and liver) content in *A. gigantea*, *H. gigas* and *P. vannamei* (unit: mmol/kg wet weight; mean ± SD). * represents significant difference ($p < 0.05$), and *** represents significant difference ($p < 0.001$).

3.2. Expression Profiling of TMAO and TMA in the Two Hadal Amphipods

In order to compare the tissue expression profiling of TMAO and TMA between the hadal amphipods and the shallow-water species, we drew a heat map regarding all the data derived from Table 1 (TMAO) and Table 2 (TMA). As shown in the heat map (Figure 3), a higher level of TMAO was detected in the external organs (such as the eye and exoskeleton) for both of the two hadal amphipods, which indicated that the TMAO concentration was not evenly distributed, although the same hydrostatic pressure existed in the outer and internal organs (Figure 3a). It also should be noticed that the TMAO content in each tissue of *A. gigantea* was higher than that of *H. gigas*, which might indicate that the TMAO content was not only correlated with the distribution depth.

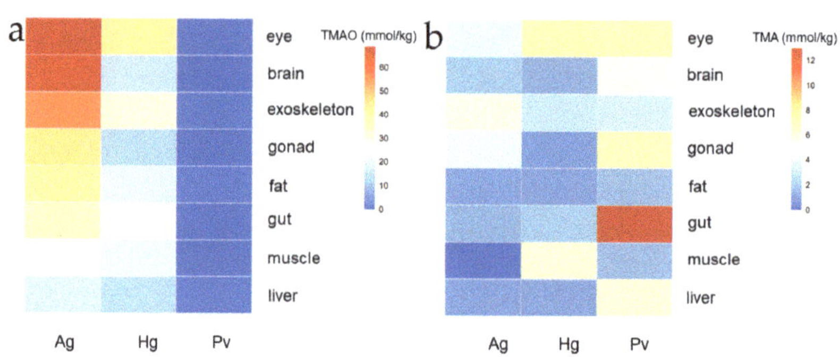

Figure 3. (**a**,**b**) TMAO and TMA concrete content in eight tissues (eye, brain, muscle, exoskeleton, gonad, fat, gut, muscle and liver) content in *A. gigantea*, *H. gigas* and *P. vannamei* (unit: mmol/kg wet weight; mean ± SD).

On the other hand, the TMA showed a different expression profiling among the three species. Generally, a similar level of TMA concentration was detected in *P. vannamei* in most of the detected tissues except for gonad, gut and liver (Figure 3b). The higher-level TMA concentrations detected in *P. vannamei* might indicate less TMA was converted to TMAO when at lower hydrostatic pressures. TMA content reached the highest level in the shallow-water *P. vannamei* gut tissue among all the three species (Figure 3b).

3.3. Correlation Analysis with TMAO and TMA Concentrations

Figure 4 showed a correlation (Pearson method) of TMA and TMAO concentrations of eight tissues in three species. There is an obvious marker that shows that in nine tissues of three species, the order of correlation coefficient between TMAO and TMA corresponding to each tissue is Hg > Ag > Pv (Figure 4a), and a strong positive correlation was found between the concentrations of TMAO and TMA in the two hadal amphipods (Figure 4b,c). It should be noticed that, with the increase in depth, the correlation is stronger (Hg > Ag). The strongest correlation in Ag is gonad, and in Hg, the strongest correlation tissue is liver (Figure 4a). However, there was no significant correlation between the tissues in the shallow-water *P. vannamei* (Figure 4d). This suggested to us that there might exist remarkable differences in the TMAO and TMA transformation processes between the hadal amphipods and shallow-water decapods.

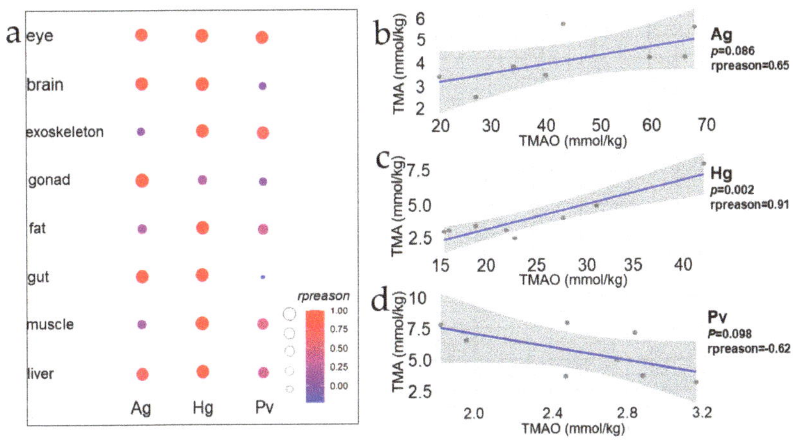

Figure 4. TMAO–TMA correlation comparison and fitting curve. (**a**) represents the TMAO-TMA correlation scatter plot of each tissue of the two hadal amphipod species, and the size and color of each point represent the magnitude of correlation coefficient. (**b–d**) represents the fitting curve of the order of *A. gigantea*, *H. gigas* and *P. vannamei*, with Pearson correlation coefficient and *p* value, respectively.

3.4. FMO3 Gene Analysis and Prediction of Three-Dimensional Crystal Structure

FMO3 is the most important TMA oxidase and converts TMA to TMAO [34,37]. In order to conduct the evolutionary analysis regarding FMO3, we based it on the phylogenetic tree constructed by the available published transcriptome data [42,43] (see Supplementary Figure S1). Supplementary Figure S2 shows the phylogenetic tree of FMO3 with the branch lengths in Gammaroidea and *A. gigantea*. As shown in Supplementary Figures S1 and S2, Gammaroidea is the closed clade to *A. gigantea*. Therefore, positive selection analysis regarding FMO3 of *A. gigantea* was performed. By setting the *A. gigantea* as the foreground branch, and other shallow-water *Gammaroidea* (*Echinogammarus marinus*, *Gammarus fossarum*, *Gammarus chevreuxi* and *Gammarus minus*) as the background species, we conducted the PSG analysis regarding *A. gigantean* FMO3. The FMO3 sequences for the background species are deduced from the SRA database (https://www.ncbi.nlm.nih.gov/sra, accessed on 5 February 2022) (*G. fossarum* ERR386132, *G. minus* SRR5576331, SRR5576333, *G. chevreuxi* SRR5109803, SRR5109804, SRR5109805, *E. marinus* SRR8089734, SRR8089735, *A. gigantea* PRJNA739006). We performed positively selected analysis by branch site model. Three positive selected sites (92S, 260A, 286 K) were found in the *A. gigantea* FMO3 gene, although likelihood ratio test (LRT) is not significant for the whole branch.

It should be noticed that some specific protein mutation sites were detected in the hadal amphipod *A. gigantea* (Supplementary Table S1). To our surprise, we found two specific sites in FAD binding motifs (GXGXXG) and FMO identifying motifs (FXGXXXHXXXF/Y) in *A. gigantea* in 9G and 176 V (Figure 5a,b). In *A. gigantea* FMO3, alanine to glycine mutations exist in FAD-binding motifs, and threonine to valine mutations exist in FMO-identifying motifs, while these two mutations do not exist in the four background species (Figure 5b), which might promote the hadal amphipod's FMO3 functions. Moreover, I-TASSER [44] was used to predict the crystal structure of the FMO3 sequence of *A. gigantea* (Figure 5a). The FAD-binding motifs and FAD ligands indicated that this mutation occurred in the conserved FDA-identifying motifs. Therefore, these positive and specific sites of *A. gigantea* FMO3 might help to adapt to the hadal environment.

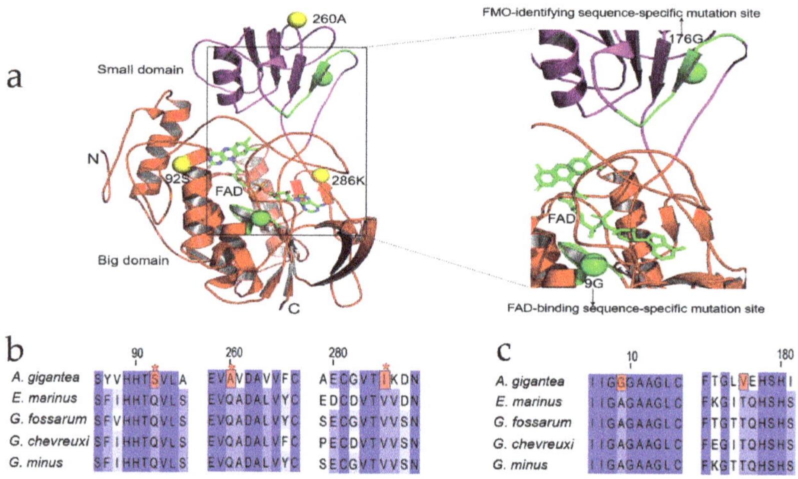

Figure 5. The three-dimensional crystal structure of FMO3 gene of *A. gigantea* and the positively selected site and specific mutation site of the FMO3 gene, (**a**) represents a three-dimensional view of the FMO3 protein, highlighting the locations of three positive selected sites (yellow spheres) and specific substitution sites on two conserved motifs (green spheres). Conserved motifs and the ligand FAD are also represented in green; (**b**) represents the location of positively selected sites in *A. gigantea*. Positively selected sites are represented in red, and * represents $p < 0.05$; (**c**) represents the locations of specific mutation sites that occur on a conserved motif in *A. gigantea* compared with the four amphipods in Gammaroidea, marked in red.

4. Discussion

The hadal zone is the deepest and most mysterious habitat on earth [45]. In hadal trench environments, amphipods appear to be vertically stratified, and species are confined to a relatively narrow depth range within each trench. Meanwhile, the high hydrostatic pressure of the hadal zone is considered to be one of the major obstacles for species to adapt to the hadal environment [46,47]. Different hadal amphipods live at specific depths, and they dominate scavenging communities and are regarded as the primary prey of hadal predators [48]. Accordingly, the study of amphipods is of great importance to the understanding of hadal environment adaptations and amphipods are often used as biological indicators of the hadal environment [49].

Based on these, in this study, we measured the TMAO and TMA concentrations of two hadal amphipods and compared with the values of the shallow-water *Penaeus vannamei*. The overall TMAO content was much higher in the hadal amphipods' tissues (Table 1, Figures 1a, 2a and 3a). Previous studies on hadal fish as well as hadal microbes have proved that HHP will gradually increase with the increase in depth, and the accumulation of TMAO in the body will gradually increase to resist high hydrostatic pressure [23,27].

However, our research shows the TMAO content in each tissue of *A. gigantea* (8824 m) was higher than that of *H. gigas* (10,839 m) (Figure 3a), which might indicate that the TMAO content is not only correlated with the distribution depth, but also related to the particular species. In fact, the two hadal amphipods were sampled from different trenches. It was reported that different TMAO concentrations were detected between the Kermadec Trench and the Mariana Trench at the same depth. However, the TMAO content increases linearly when in the same environment [30]. Since the TMAO is produced in cell and the dead animals will reduce TMAO to TMA [50], it is possible that *A. gigantea* tissues have more intracellular and less extracellular space than tissues of *H. gigas*; therefore, *H. gigas* could have more TMAO inside cells than *A. gigantea*, and it should not be overlooked how long these specimens suffered until they were pulled to the surface. Compared with *A. gigantean* (8824 m), it took a longer time to collect *H. gigas* (10,839 m) samples, which suggested that more TMAO might be converted to TMA in *H. gigas*. The TMA data happened to show that TMA content is higher in *H. gigas* than in *A. gigantea* (Figure 3b).

It should be noticed that a higher level of TMAO was detected in the external organs (such as eye and exoskeleton) for both of the two hadal amphipods (Figure 3a). Previous studies have shown that TMAO can over-stabilize proteins and reduce protein activities and functions in an excessive concentration [51]. However, the normal activities of the hadal species are mainly determined by tissues, internal organs and proteases. There may exist a transport mechanism that transport TMAO in internal organs to external organs to keep the TMAO in a balanced concentration.

TMA has been reported to be produced by gut microbes in hadal fishes [52]. Different depths and the gut microbe's diversity in different amphipods are different [53], and the available nitrogen sources may decrease as the depth increases. TMA production is also affected by the interaction between intestinal microorganisms and hadal amphipod species [3]. As shown in Figure 3b, the tissues with the highest TMA content are found in the intestine of *P. vannamei*. It was well known that the liver is considered to be the main site of TMAO production in mammals [54]. The highest TMA content located in *P. vannamei*'s gut revealed that the nitrogen source is sufficient in *P. vannamei* and TMA in crustaceans might also be produced by gut microbes. The *P. vannamei* samples were captured in Qingdao, but during the transporting process to Shanghai, unfortunately it was not possible to keep it constantly at −80 °C. In transit, the TMA in *P. vannamei* probably started to convert TMA to TMAO in the gut by bacteria [55].

The muscle TMAO content is nearly 400 mmol/kg in snailfish when the depth is nearly 8000 m and is nearly 200 mmol/kg in deep-sea decapods when the depth is nearly 2000 m [29]. However, our data showed the highest TMAO content in eight tissues is 68 mmol/kg in hadal amphipods and it is obviously lower than those two species. At the same time, the TMAO content has been reported to exist in the cell [50]. The tissue samples used in our study may contain non-cell parts and the cell structure may also have been broken during the tissue dissection process, which could be the possible reason for the lower TMAO concentrations detected in our study.

Not only TMAO, but also TMA, plays an important role in marine ecosystems as a major precursor of the greenhouse gas methane in coastal sediments [56]. It has also been demonstrated that microorganisms in the deep ocean absorb TMA from the environment and convert it to TMAO to offset HHP [27]. Our study shows a strong positive correlation between the two hadal amphipod TMAO–TMA species in deeper waters, but this correlation was not confirmed in the shallow-water *P. vannamei*.

It was well known that the conversion of TMA to TMAO is accomplished through the FMO3 gene. The FMO gene has also been shown to prolong life and maintain structural stability [57]. Therefore, the study of amphipods may reveal many phenomena of the hadal environment and adaptive evolution of species. In this study, we used the hadal amphipod *A. gigantea's* FMO3 gene to analyze the selected pressure of the hadal amphipod species and found that FMO3 sequences generally did not have the positive selection in the entire nucleic acid evolutionary branches of hadal species. Structural adaptations of proteins to hadal conditions may include patterns of amino acid substitution and changes in protein structure that counteract the effects of stress on protein function, and even evolutionary patterns of some proteins that respond to hydrostatic pressure [7,58]. Indeed, Mariana Trench snailfish (MHS) also exhibit a positive selected site in FMO. It not even have a closed skull as the bone Gla protein (*bglap*) gene has a frameshift mutation that may cause the premature termination of cartilage calcification in the MHS [34].

Therefore, proteins at some sites are positively selected to adapt to high hydrostatic pressure to maintain structural stability. The FMO3 was observed in two FAD-specific binding motifs in the *A. gigantea*, and two FMO3-conserved motifs are mutated at specific sites. Therefore, the substitution of such a conserved site suggests that this mutation may be functionally beneficial to the protein functional adaptation of hadal species to hadal environments. FMO3 is an evolutionarily conserved and highly abundant redox enzyme system. The crystal structure of the FMO3 protein of *A. gigantea* was predicted by the threading method as shown in Supplementary Figure S3. Factually, the genomic gene family expansion, amino acid substitution and copy number increase are also important regulatory mechanisms. Like HSPs in the hadal amphipod and fish, the HSP in *H. gigas* can have more copies and some special site mutations. Other species help adapt HHP; the λmax of rhodopsin 1 in YHS (Yap hadal snailfish) and MHS rhodopsin are lower than the levels found in shallow-water teleosts and may help them adapt to dark environments in the hadal zone [34,59,60]. All in all, the concrete intrinsic adaptation is complicated. Further studies of specific structure and function will clarify the specific effects of this unique mutation on the FMO3 protein. The concrete mechanism about how the hadal species adapt to the hadal environment is still to be researched.

5. Conclusions

In this study, we measured the TMAO and TMA content in eight tissues of two hadal amphipod species, and a shallow water decapod, *P. vannamei*. We found that the TMAO content in hadal amphipod species is much higher than that in shallow species, and also found that in amphipods, there may be a strong positive TMAO–TMA correlation among amphipods in hadal habitats. We also found specific loci changes on two conserved motifs in the FMO3 gene of the hadal amphipod. This study provides insights into the molecular adaptation mechanisms of hadal organisms.

Supplementary Materials: The following supporting information can be downloaded at: https://www.mdpi.com/article/10.3390/jmse10040454/s1. Figure S1: The phylogenetic position of the *Hirondellea gigas* and *Alicella gigantea* based on maximum likelihood (ML) estimation of FMO3 gene of selected crustacean species. Figure S2: Phylogenetic tree of FMO3 with the branch lengths between Gammaroidea and *A. gigantea*. Figure S3: The prediction of three-dimensional crystal structure of FMO3 protein of *A. gigantea*; Table S1: The positive selected sites and specific substitution sites on two conserved motifs.

Author Contributions: Validation, Q.L., S.J. and W.L.; formal analysis, Q.L.; data curation, Q.L., S.J. and W.L.; sampling, B.P.; writing—original draft preparation, Q.L. and S.J.; writing—review and editing, Q.X.; visualization, Q.L.; supervision, Q.X.; project administration, Q.X.; funding acquisition, Q.X. All authors have read and agreed to the published version of the manuscript.

Funding: This research was funded in part by the Funding Project of the National Key Research and Development Program of China (2018YFC0310600), the National Key Research and Development Program of China (2018YFD0900601), the National Natural Science Foundation of China (Grant No. 31772826), and the major scientific innovation project from Shanghai Committee of Education (2017-01-07-00-10-E00060).

Institutional Review Board Statement: Not applicable.

Informed Consent Statement: Not applicable.

Data Availability Statement: Not applicable.

Acknowledgments: We would like to thank Shanghai Rainbowfish Ocean Technology Co., Ltd., for sample collection. We also thank Weicheng Cui's and Jiasong Fang's research group members and other people for sample collection.

Conflicts of Interest: The authors declare no conflict of interest.

References

1. Cressey, D. The Hadal Zone: Life in the Deepest Oceans. *Nature* **2015**, *523*, SB1.
2. Weston, J.N.J.; Espinosa-Leal, L.; Wainwright, J.A.; Stewart, E.C.D.; Gonzalez, C.E.; Linley, T.D.; Reid, W.D.K.; Hidalgo, P.; Oliva, M.E.; Ulloa, O.; et al. *Eurythenes atacamensis* sp. nov. (Crustacea: Amphipoda) exhibits ontogenetic vertical stratification across abyssal and hadal depths in the Atacama Trench, eastern South Pacific Ocean. *Mar. Biodivers.* **2021**, *51*, 51. [CrossRef] [PubMed]
3. Chan, J.L.; Geng, D.Q.; Pan, B.B.; Zhang, Q.M.; Xu, Q.H. Metagenomic Insights Into the Structure and Function of Intestinal Microbiota of the Hadal Amphipods. *Front. Microbiol.* **2021**, *12*. Available online: https://www.frontiersin.org/articles/10.3389/fmicb.2021.668989/full (accessed on 5 February 2022). [CrossRef]
4. Zhang, W.P.; Tian, R.M.; Sun, J.; Bougouffa, S.; Ding, W.; Cai, L.; Lan, Y.; Tong, H.Y.; Li, Y.X.; Jamieson, A.J.; et al. Genome Reduction in Psychromonas Species within the Gut of an Amphipod from the Ocean's Deepest Point. *Msystems* **2018**, *3*, e00009-18. [CrossRef] [PubMed]
5. Bartlett, D.H. Microbial life at high pressures. *Sci. Prog.* **1992**, *76*, 479–496. [PubMed]
6. Jamieson, A.J.; Fujii, T.; Mayor, D.J.; Solan, M.; Priede, I.G. Hadal trenches: The ecology of the deepest places on Earth. *Trends Ecol. Evol.* **2010**, *25*, 190–197. [CrossRef] [PubMed]
7. Somero, G.N. Adaptations to high hydrostatic pressure. *Annu. Rev. Physiol.* **1992**, *54*, 557–577. [CrossRef] [PubMed]
8. Macdonald, A.G. Hydrostatic pressure as an environmental factor in life processes. *Comp. Biochem. Phys. A* **1997**, *116*, 291–297. [CrossRef]
9. Thurston, M.H. Scavenging abyssal amphipods from the North-East Atlantic ocean. *Mar. Biol.* **1979**, *51*, 55–68. [CrossRef]
10. Blankenship, L.E.; Yayanos, A.A.; Cadien, D.B.; Levin, L.A. Vertical zonation patterns of scavenging amphipods from the Hadal zone of the Tonga and Kermadec Trenches. *Deep Sea Res. Part I Oceanogr. Res. Pap.* **2006**, *53*, 48–61. [CrossRef]
11. Beliaev, G.M.; Vinogradov, M.Y. Deep-sea ocean trenches and their fauna. *Scripps Inst. Oceanogr.* **1989**. Available online: https://escholarship.org/uc/item/46n6148x (accessed on 5 February 2022).
12. Jamieson, A.J.; Lacey, N.C.; Lorz, A.N.; Rowden, A.A.; Piertney, S.B. The supergiant amphipod Alicella gigantea (Crustacea: Alicellidae) from hadal depths in the Kermadec Trench, SW Pacific Ocean. *Deep Sea Res. Part II Top. Stud. Oceanogr.* **2013**, *92*, 107–113. [CrossRef]
13. Zhu, L.Y.; Geng, D.Q.; Pan, B.B.; Li, W.H.; Jiang, S.W.; Xu, Q.H. Trace Elemental Analysis of the Exoskeleton, Leg Muscle, and Gut of Three Hadal Amphipods. *Biol. Trace Elem. Res.* **2021**, *200*, 1395–1407. [CrossRef] [PubMed]
14. Kobayashi, H.; Hatada, Y.; Tsubouchi, T.; Nagahama, T.; Takami, H. The Hadal Amphipod Hirondellea gigas Possessing a Unique Cellulase for Digesting Wooden Debris Buried in the Deepest Seafloor. *PLoS ONE* **2012**, *7*, e42727. [CrossRef] [PubMed]
15. Silva, J.L.; Vieira, T.C.R.G.; Gomes, M.P.B.; Bom, P.A.; Lima, L.M.T.R.; Freitas, M.S.; Ishimaru, D.; Cordeiro, Y.; Foguel, D. Ligand Binding and Hydration in Protein Misfolding: Insights from Studies of Prion and p53 Tumor Suppressor Proteins. *Acc. Chem. Res.* **2010**, *43*, 271–279. [CrossRef] [PubMed]
16. Lan, Y.; Sun, J.; Xu, T.; Chen, C.; Tian, R.M.; Qiu, J.W.; Qian, P.Y. De novo transcriptome assembly and positive selection analysis of an individual deep-sea fish. *BMC Genom.* **2018**, *19*, 1–9. [CrossRef]
17. Brooks, N.J. Pressure effects on lipids and bio-membrane assemblies. *IUCrJ* **2014**, *1*, 470–477. [CrossRef]
18. Somero, G.N. Protein adaptations to temperature and pressure: Complementary roles of adaptive changes in amino acid sequence and internal milieu. *Comp. Biochem. Phys. B* **2003**, *136*, 577–591. [CrossRef]
19. Morris, J.P.; Thatje, S.; Hauton, C. The use of stress-70 proteins in physiology: A re-appraisal. *Mol. Ecol.* **2013**, *22*, 1494–1502. [CrossRef]
20. Genest, O.; Wickner, S.; Doyle, S.M. Hsp90 and Hsp70 chaperones: Collaborators in protein remodeling. *J. Biol. Chem.* **2019**, *294*, 2109–2120. [CrossRef]
21. Yancey, P.H. Organic osmolytes as compatible, metabolic and counteracting cytoprotectants in high osmolarity and other stresses. *J. Exp. Biol.* **2005**, *208*, 2819–2830. [CrossRef]
22. Wang, Z.; Klipfell, E.; Bennett, B.J.; Koeth, R.; Levison, B.S.; Dugar, B.; Feldstein, A.E.; Britt, E.B.; Fu, X.; Chung, Y.M.; et al. Gut flora metabolism of phosphatidylcholine promotes cardiovascular disease. *Nature* **2011**, *472*, 57–63. [CrossRef] [PubMed]
23. Canyelles, M.; Tondo, M.; Cedo, L.; Farras, M.; Escola-Gil, J.C.; Blanco-Vaca, F. Trimethylamine N-Oxide: A Link among Diet, Gut Microbiota, Gene Regulation of Liver and Intestine Cholesterol Homeostasis and HDL Function. *Int. J. Mol. Sci.* **2018**, *19*, 3228. [CrossRef] [PubMed]
24. Jamieson, D.A. Marine fish may be biochemically constrained from inhabiting deepest ocean depths. *Proc. Natl. Acad. Sci. USA* **2014**, *111*, 4461–4465.
25. Yancey, P.H. Cellular responses in marine animals to hydrostatic pressure. *J. Exp. Zool. Part A* **2020**, *333*, 398–420. [CrossRef] [PubMed]

26. Bolen, D.W.; Baskakov, I.V. The osmophobic effect: Natural selection of a thermodynamic force in protein folding. *J. Mol. Biol.* **2001**, *310*, 955–963. [CrossRef]
27. Qin, Q.L.; Wang, Z.B.; Su, H.N.; Chen, X.L.; Miao, J.; Wang, X.J.; Li, C.Y.; Zhang, X.Y.; Li, P.Y.; Wang, M.; et al. Oxidation of trimethylamine to trimethylamine N-oxide facilitates high hydrostatic pressure tolerance in a generalist bacterial lineage. *Sci. Adv.* **2021**, *7*, eabf9941. [CrossRef]
28. Yancey, P.H.; Speers-Roesch, B.; Atchinson, S.; Reist, J.D.; Majewski, A.R.; Treberg, J.R. Osmolyte Adjustments as a Pressure Adaptation in Deep-Sea Chondrichthyan Fishes: An Intraspecific Test in Arctic Skates (*Amblyraja hyperborea*) along a Depth Gradient. *Physiol. Biochem. Zool.* **2018**, *91*, 788–796. [CrossRef]
29. Kelly, R.H.; Yancey, P.H. High Contents of Trimethylamine Oxide Correlating with Depth in Deep-Sea Teleost Fishes, Skates, and Decapod Crustaceans. *Biol. Bull.* **1999**, *196*, 1825. [CrossRef]
30. Downing, A.B.; Wallace, G.T.; Yancey, P.H. Organic osmolytes of amphipods from littoral to hadal zones: Increases with depth in trimethylamine N-oxide, scyllo-inositol and other potential pressure counteractants. *Deep Sea Res. Part I Oceanogr. Res. Pap.* **2018**, *138*, 1–10. [CrossRef]
31. Linley, T.D.; Gerringer, M.E.; Yancey, P.H.; Drazen, J.C.; Weinstock, C.L.; Jamieson, A.J. Fishes of the hadal zone including new species, in situ observations and depth records of Liparidae. *Deep-Sea Res. Part I* **2016**, *114*, 99–110. [CrossRef]
32. Swan, J.A.; Jamieson, A.J.; Linley, T.D.; Yancey, P.H. Worldwide distribution and depth limits of decapod crustaceans (Penaeoidea, Oplophoroidea) across the abyssal-hadal transition zone of eleven subduction trenches and five additional deep-sea features. *J. Crustacean Biol.* **2021**, *41*, ruaa102. [CrossRef]
33. Winnikoff, J.R.; Wilson, T.M.; Thuesen, E.V.; Haddock, S.H.D. Enzymes feel the squeeze: Biochemical adaptation to pressure in the deep sea. *Biochemist* **2017**, *39*, 26–29. [CrossRef]
34. Wang, K.; Shen, Y.; Yang, Y.; Gan, X.; Liu, G.; Hu, K.; Li, Y.; Gao, Z.; Zhu, L.; Yan, G. Morphology and genome of a snailfish from the Mariana Trench provide insights into deep-sea adaptation. *Nat. Ecol. Evol.* **2019**, *3*, 823–833. [CrossRef] [PubMed]
35. Van Berkel, W.J.H.; Kamerbeek, N.M.; Fraaije, M.W. Flavoprotein monooxygenases, a diverse class of oxidative biocatalysts. *J. Biotechnol.* **2006**, *124*, 670–689. [CrossRef]
36. Eswaramoorthy, S.; Bonanno, J.B.; Burley, S.K.; Swaminathan, S. Mechanism of action of a flavin-containing monooxygenase. *Proc. Natl. Acad. Sci. USA* **2006**, *103*, 9832–9837. [CrossRef]
37. Huang, S.; Howington, M.B.; Dobry, C.J.; Evans, C.R.; Leiser, S.F. Flavin-Containing Monooxygenases Are Conserved Regulators of Stress Resistance and Metabolism. *Front. Cell Dev. Biol.* **2021**, *9*, 630188. [CrossRef]
38. Beaty, N.B.; Ballou, D.P. The oxidative half-reaction of liver microsomal FAD-containing monooxygenase. *J. Biol. Chem.* **1981**, *256*, 4619–4625. [CrossRef]
39. Wu, Q.; Zhao, Y.; Zhang, X.; Yang, X. A faster and simpler UPLC-MS/MS method for the simultaneous determination of trimethylamine N-oxide, trimethylamine and dimethylamine in different types of biological samples. *Food Funct.* **2019**, *10*, 6484–6491. [CrossRef]
40. Gao, F.; Chen, C.; Arab, D.A.; Du, Z.; He, Y.; Ho, S.Y.W. EasyCodeML: A visual tool for analysis of selection using CodeML. *Ecol. Evol.* **2019**, *9*, 3891–3898. [CrossRef]
41. Zerbst-Boroffka, I.; Kamaltynow, R.M.; Harjes, S.; Kinne-Saffran, E.; Gross, J. TMAO and other organic osmolytes in the muscles of amphipods (Crustacea) from shallow and deep water of Lake Baikal. *Comp. Biochem. Physiol. Part A Mol. Integr. Physiol.* **2005**, *142*, 58–64. [CrossRef]
42. Li, W.H.; Wang, F.X.; Jiang, S.W.; Pan, B.B.; Chan, J.L.; Xu, Q.H. The Adaptive Evolution and Gigantism Mechanisms of the Hadal "Supergiant" Amphipod Alicella gigantea. *Front. Mar. Sci.* **2021**, *8*. Available online: https://www.frontiersin.org/articles/10.3389/fmars.2021.743663/full (accessed on 5 February 2022). [CrossRef]
43. Copilas-Ciocianu, D.; Borko, S.; Fiser, C. The late blooming amphipods: Global change promoted post-Jurassic ecological radiation despite Palaeozoic origin. *Mol. Phylogenet. Evol.* **2020**, *143*, 106664. [CrossRef] [PubMed]
44. Zheng, W.; Zhang, C.; Li, Y.; Pearce, R.; Zhang, Y. Folding non-homologous proteins by coupling deep-learning contact maps with I-TASSER assembly simulations. *Cell Rep. Methods* **2021**, *1*, 100014. [CrossRef]
45. Piccard, J. Man's deepest dive. *Natl. Geogr.* **1960**, *118*, 224–239.
46. Brown, A.; Thatje, S. Explaining bathymetric diversity patterns in marine benthic invertebrates and demersal fishes: Physiological contributions to adaptation of life at depth. *Biol. Rev.* **2013**, *89*, 406–426. [CrossRef] [PubMed]
47. Smith, K.E.; Brown, A.; Thatje, S. The metabolic cost of developing under hydrostatic pressure: Experimental evidence supports macroecological pattern. *Mar. Ecol. Prog. Ser.* **2015**, *524*, 71–82. [CrossRef]
48. Jamieson, A.J.; Fujii, T.; Solan, M.; Matsumoto, A.K.; Bagley, P.M.; Priede, I.G. First findings of decapod crustacea in the hadal zone. *Deep Sea Res. Part I Oceanogr. Res. Pap.* **2009**, *56*, 641–647. [CrossRef]
49. Hupalo, K.; Teixeira, M.A.L.; Rewicz, T.; Sezgin, M.; Iannilli, V.; Karaman, G.S.; Grabowski, M.; Costa, F.O. Persistence of phylogeographic footprints helps to understand cryptic diversity detected in two marine amphipods widespread in the Mediterranean basin. *Mol. Phylogenet. Evol.* **2019**, *132*, 53–66. [CrossRef]
50. Gillett, M.B.; Suko, J.R.; Santoso, F.O.; Yancey, P.H. Elevated levels of trimethylamine oxide in muscles of deep-sea gadiform teleosts: A high-pressure adaptation? *J. Exp. Zool.* **1997**, *279*, 386–391. [CrossRef]
51. Yancey, P.H.; Clark, M.E.; Hand, S.C.; Bowlus, R.D.; Somero, G.N. Living with water stress: Evolution of osmolyte systems. *Science* **1982**, *217*, 1214–1222. [CrossRef]

52. Subramaniam, S.; Fletcher, C. Trimethylamine N-oxide: Breathe new life. *Br. J. Pharmacol.* **2018**, *175*, 1344–1353. [CrossRef] [PubMed]
53. Chan, J.; Geng, D.; Pan, B.; Zhang, Q.; Xu, Q. Gut Microbial Divergence Between Three Hadal Amphipod Species from the Isolated Hadal Trenches. *Microb. Ecol.* **2021**, 1–11. [CrossRef]
54. Janeiro, M.H.; Ramirez, M.J.; Milagro, F.I.; Martinez, J.A.; Solas, M. Implication of Trimethylamine N-Oxide (TMAO) in Disease: Potential Biomarker or New Therapeutic Target. *Nutrients* **2018**, *10*, 1398. [CrossRef] [PubMed]
55. Zhang, W.Q.; Wang, Y.J.; Zhang, A.; Ding, Y.J.; Zhang, X.N.; Jia, Q.J.; Zhu, Y.P.; Li, Y.Y.; Lv, S.C.; Zhang, J.P. TMA/TMAO in Hypertension: Novel Horizons and Potential Therapies. *J. Cardiovasc. Transl. Res.* **2021**, *14*, 1117–1124. [CrossRef]
56. King, G.M. Metabolism of trimethylamine, choline, and glycine betaine by sulfate-reducing and methanogenic bacteria in marine sediments. *Appl. Environ. Microbiol.* **1984**, *48*, 719–725. [CrossRef]
57. Leiser, S.F.; Miller, H.; Rossner, R.; Fletcher, M.; Leonard, A.; Primitivo, M.; Rintala, N.; Ramos, F.J.; Miller, D.L.; Kaeberlein, M. Cell nonautonomous activation of flavin-containing monooxygenase promotes longevity and health span. *Science* **2015**, *350*, 1375–1378. [CrossRef]
58. Yafremava, L.S.; Di Giulio, M.; Caetano-Anolles, G. Comparative analysis of barophily-related amino acid content in protein domains of Pyrococcus abyssi and Pyrococcus furiosus. *Archaea* **2013**, *2013*, 680436. [CrossRef]
59. Ritchie, H.; Jamieson, A.J.; Piertney, S.B. Heat-shock protein adaptation in abyssal and hadal amphipods. *Deep Sea Res. Part II Top. Stud. Oceanogr.* **2018**, *155*, 61–69. [CrossRef]
60. Mu, Y.; Bian, C.; Liu, R.; Wang, Y.; Shao, G.; Li, J.; Qiu, Y.; He, T.; Li, W.; Ao, J.; et al. Whole genome sequencing of a snailfish from the Yap Trench (~7000 m) clarifies the molecular mechanisms underlying adaptation to the deep sea. *PLoS Genet.* **2021**, *17*, e1009530. [CrossRef] [PubMed]

Journal of
Marine Science and Engineering

MDPI

Article

The Symbiotic Relationship between the Antarctic Limpet, *Nacella concinna*, and Epibiont Coralline Algae

Boongho Cho [1,2,†], Hyeonmi Bae [1,2,†] and Taewon Kim [1,2,*]

1 Department of Ocean Sciences, Inha University, 100 Inha-ro, Michuhol-gu, Incheon 22212, Korea; boonghocho@gmail.com (B.C.); qogusala@gmail.com (H.B.)
2 Program in Biomedical Science and Engineering, Inha University, 100 Inha-ro, Michuhol-gu, Incheon 22212, Korea
* Correspondence: ktwon@inha.ac.kr; Tel.: +82-10-8726-3070
† These authors contributed equally to this work.

Abstract: The Antarctic limpet, *Nacella concinna*, is one of the most abundant benthic marine invertebrates found in the intertidal zone of King George Island, Antarctica. The shell of *N. concinna* is often encrusted with the coralline algae *Clathromorphum obtectulum*. In this study, to reveal the relationship between the limpet and coralline algae, we examined how the coralline algae affect the physical condition (survival and health) and morphology of the limpet. We cultured the limpets for 22 days and compared mortality, weight, condition factor (CF), fatty acid content, and the structure of the shell surface between limpets both with and without coralline algae in the laboratory. We also measured the environmental factors (i.e., temperature, pH, and salinity) of the seawater at each sampling site and the CF of the limpets and correlated them with coverage of coralline algae. The presence of coralline algae significantly increased the mortality of the limpets by 40% and the shell weight by 1.4-fold but did not affect the CF. Additionally, coralline algae altered the fatty acid profiles related to the limpet's lipid metabolism (saturated fatty acids (SFA) and some polyunsaturated fatty acids (PUFA)). Specifically, C16:0, C17:0, C18:0, and total SFA increased, whereas C18:2 and C18:3 decreased. However, observations with a scanning electron microscope showed that shell damage in limpets with coralline algae was much less than in limpets without coralline algae, suggesting that coralline algae may provide protection against endolithic algae. The area of coralline algae on the limpet shell was positively correlated with the pH and temperature of the seawater. The results suggest that although coralline algae are generally assumed to be parasitical, the relationship between *N. concinna* and coralline algae may change to mutualism under certain conditions.

Keywords: epibiosis; antarctica; limpet; coralline algae; parasite

1. Introduction

Epibiosis is a phenomenon that is common in marine benthos [1,2], where the basibiont acts as the substrate to which a sessile epibiont can attach [2]. Numerous epibiotic studies have been conducted [3–5]. For example, Manning and Lindquist [6] studied the interaction between clams and hydroids on sandy beaches and found that the hydroid, as the epibiont, adversely affected the movement of the clams. Marin and Belluga [7] analyzed whether epibiotic sponges protected clams against predation and found that bivalves fouled with sponges survived significantly longer than unfouled bivalves. Wahl and Hay [8] examined the relationship between seaweeds, epibiotic plants, and the predatory sea urchin. They found that the sea urchins' preference for seaweed changed according to the presence or absence of epibionts. However, thus far, there have only been a few studies of epibiosis in more extreme environments [9–11].

Antarctica is one of the most sensitive regions of the world to climate change [12,13], and research on ocean acidification and ocean freshening in the area is ongoing [14–17]. The Antarctic limpet, *Nacella concinna*, is one of the most conspicuous of the macrobenthos in the rocky intertidal and subtidal zones of Antarctica [18]. The shell of the limpet (the basibiont) can provide a habitat for the epibiont, and N. concinna that were fouled with coralline algae (epibiont) are commonly found [19,20]. The coralline algae, *Clathromorphum obtectulum,* is one of the most abundant non-geniculate coralline algae found throughout the Arctic and Antarctic regions and live on the rigid substrates and shells of various organisms [21–24]. McClintock et al. [19] showed that coralline algae can produce Mg-calcite skeletons when they cover the shell of *N. concinna*, providing protection to the shell in low pH environments. Schoenrock et al. [22] conducted experiments on culturing calcified species (*C. obtectulum*) and Antarctic crustose macroalgae (*Hildenbrandia* sp.) under the conditions of increased seawater temperature and pCO_2 according to near-future climate change. They reported *Hildenbrandia* sp. would have potential competitive advantages for intertidal space. However, until now, there have been no studies on how epibiotic coralline algae directly affects the limpet itself.

In this study, we aimed to understand the relationship between the limpet (*N. concinna*) and coralline algae (*C. obtectulum*) in Antarctica for the first time. We hypothesized that epibiotic coralline algae on the shell would adversely affect the limpet. Accordingly, we predicted that the coralline algae would decrease the survival condition factor (CF), increase shell weight, and change the fatty acid content in the tissue of the limpet.

2. Materials and Methods

2.1. Sampling (Culture and Field Survey)

The limpets (*Nacella concinna*) for the culture experiments were collected in 12 January 2019 by hand from the outer side of Marian Cove on King George Island (latitude 62°14.531′ S, longitude 58°44.783′ W; water temperature: 2.8 °C ± 0.2, salinity: 33.3 psu ± 0.2, pH: 8.3 ± 0.004, mean ± standard error (SE); Figure 1). The limpets were selected based on shell length (32.7 ± 0.2, mean ± SE). To determine the area of coralline algae on the shells of the limpets, we analyzed the images (camera: Olympus TG-5, Olympus, Tokyo, Japan) using ImageJ®. We then divided the limpets into groups depending on whether they had more than 10% coralline algae coverage (present) or not (absent) (N = 20 per each treatment).

(**a**) (**b**)

Figure 1. (**a**) The sampling site for *Nacella concinna* is marked by a white circle (culture experiments) and white triangles (field survey). The insert on the upper left indicates the location of Marian Cove (red square) on King George Island. (**b**) Sample images of *N. concinna* were used in this study. The upper row shows *N. concinna* partially covered by the coralline algae *Clathromorphum obtectulum*. The lower row shows *N. concinna* without *C. obtectulum* on the shell.

The limpets ($N = 69$) for the field survey for determining the relationships between the limpets and environmental factors (pH, temperature, and salinity) were collected from 19 different points both inside and outside the ice wall from 25 December 2018, to 8 February 2019 (Figure 1). Each sampling was conducted within a 30 cm radius of a sampling spot. The temperature, salinity, and dissolved oxygen content were measured using multiple water quality sensors (YSI pro2030, Yellow Spring Instruments Inc., Yellow Springs, OH, USA), while pH was measured with a pH meter (Seven2Go pH/Ion meter S8, Mettler Toledo, Columbus, OH, USA).

2.2. Experiment Setup and Acclimation

Acclimatizing the limpets to the experimental conditions was conducted in a water tank (80 × 45 × 20 cm) using seawater (pH 8.0, salinity 34 psu) taken from in front of King Sejong Station over seven days from 12 to 19 January 2019. To control salinity, the frozen part was melted for use in low-saline seawater after freezing, and the rest was used for high-saline seawater [17]. pH was controlled using CO_2 tablets (SERA, Heinsberg, Germany) [17,25].

The culture experiment was conducted over 22 days from 19 January to 9 February 2019. The limpets were divided into two groups according to the presence or absence of coralline algae on their shells. The limpets were positioned separately, and the seawater was replaced every 24 h to maintain water quality. To maintain experimental accuracy, the top of the beaker was sealed by parafilm, and there was no food supply during the experiment. We used a low-temperature incubator (Plant Growth Chamber SH-303, Seyoung Scientific CO., Bucheon, Korea) to keep a stable temperature and humidity. The temperature, salinity, dissolved oxygen content, and pH were measured once every 24 h.

2.3. Analysis Method

During the experiment, limpet mortality was checked every 12 h and any dead limpets were immediately removed. Mortality was checked by foot muscle and tentacle movements [25]. Digital calipers (CD-15PSX, Mitutoyo Corp., Kanagawa, Japan) were used for measuring the length, height, and width of the limpets, and an electronic micro-scale (PG2002-S, Mettler Toledo, Columbus, OH, USA) was used to measure wet weight.

The CF was used as a bioindicator to measure limpet health [26,27]. After the experiment, CFs were calculated with only live individuals using the following formula [25].

$$CF = \text{tissue wet mass} \times \text{shell volume}^{-1}$$

Fatty acid analysis was performed on the foot tissue of the limpet ($N = 20$ per each treatment). The extraction of fatty acid methyl esters (FAMEs) was performed according to the methodology of Gracés and Mancha [28]. Twenty limpets were used for each treatment. The tissue samples were freeze-dried at −95 °C using freeze dryers (CoolSafe 4–15 L, LaboGene, Lillerød, Denmark). All samples were then stored at −20 °C [29,30] before being crushed and placed in tubes with Teflon caps.

The samples were mixed 2 mL of a methylation mixture (MeOH: Benzene: DMP (2,2-Dimethoxy-propane): H_2SO_4, 39:20:5:2) and 1 mL of heptane. The samples were extracted at 80 °C for two hours. After heating, the samples were cooled at room temperature before being divided into two layers. The upper layer of each sample was transferred to a vial for gas chromatography (GC) injection. The fatty acid composition was analyzed using a Gas Chromatograph (GC; Agilent 7890 A, Santa Clara, CA, USA) equipped with a 120 mm × 0.25 mm × 0.25 μm capillary column (DB-23, Agilent, Santa Clara, CA, USA) and a flame ionization detector (FID). The injector temperature was 250 °C. The fatty acid content was calculated using the internal injection standard (C15:0) of known concentration. The results were expressed as mg FA/g lipid.

Using field emission-scanning electronic microscopy (FE-SEM; S-4300SE, Hitachi, Ltd., Tokyo, Japan), species identification of the coralline algae on the shell was conducted based on the shape ($N = 1$; Figure 2a). The cross section of the shell was then analyzed to identify the direct effect of coralline algae on the limpet shell (Figure 2b).

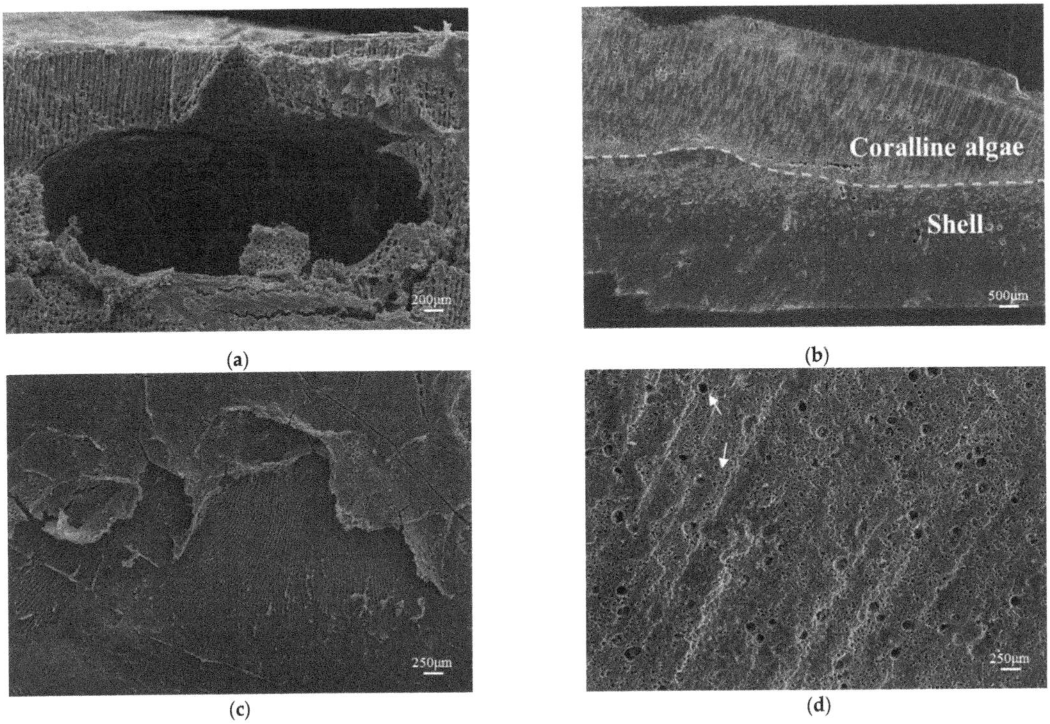

Figure 2. (**a**) SEM images of the conceptacles of coralline algae (×250), (**b**) the cross section of the limpet shell with coralline algae (×100), (**c**) the in-plane section of the limpet shell after the coralline algae had been removed deliberately (×180), and (**d**) the in-plane section of limpet shell with no coralline algae (×180). The dashed yellow line indicates the boundary line between the shell and coralline algae and the white arrows indicate irregular holes.

2.4. Statistical Analysis

We tested for significant differences in mortality, CF, shell weight, coverage, and fatty acid content using the two-tailed independent *t*-test for normally distributed data and the Mann–Whitney *U* test for non-normally distributed data. Significance probability levels of the fatty acid analysis were recalculated using the sequential Bonferroni correction for multiple comparisons [31]. We used Spearman's rank correlation coefficient to investigate the relationship between coralline algae coverage on the limpet shells and environmental factors (pH, salinity, and temperature) or the CF of the limpet. All statistical analyses were performed using SPSS software (version 19.0; SPSS, Inc., Armonk, NY, USA), with a *p*-value of ≤ 0.05 denoting statistical significance.

3. Results

3.1. Mortality, Shell Weight, and CF

Mortality rate (%) and shell weight (g) was significantly higher in limpets with coralline algae on their shell (50% and 1.99 ± 0.10 g) compared to limpets without coralline algae on their shell (10% and 1.44 ± 0.071 g) (Two-tailed Mann–Whitney *U* test; $U = 120$, $n_1 = 20$, $n_2 = 20$, $p = 0.006$; Figure 1a; Two-tailed independent *t*-test, $t = 4.331$, $df = 38$, $p < 0.001$; Figure 3a,b). However, there was no significant difference in the CF between the two groups (Two-tailed independent t-test, $t = -0.389$, $df = 26$, $p = 0.700$; Figure 3c). Additionally, the area of algae coverage (%) did not significantly influence mortality (Two-tailed independent t-test, $t = 1.447$, $df = 18$, $p = 0.165$; Figure 3d).

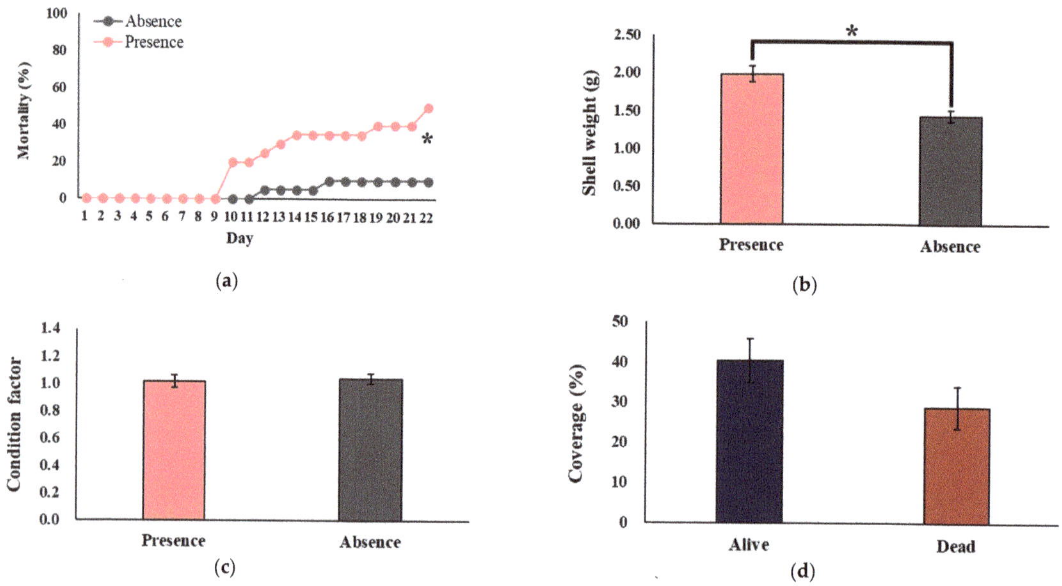

Figure 3. (**a**) Mortality, (**b**) shell weight, and (**c**) CF of limpets in the presence or absence of coralline algae on the shell. (**d**) The coverage area (%) of the coralline algae from alive or dead limpets within the group in which coralline algae are present on the shell. Values for shell weight, CF, and coverage area are indicated as mean ± SE. Asterisks indicate significant differences between groups ($p < 0.05$).

3.2. Fatty Acid Content

As a result of analyzing the fatty acids of two groups with the presence or absence of coralline algae, significant differences were found in SFA among the main groups, SFA, MUFA, and PUFA (Table 1). The group in which coralline algae was present on the shell had a higher content of SFA (C16:0, C17:0, C18:0) and lower content of PUFA (C18:2, C18:3) than that of the group without coralline algae.

Table 1. Fatty acids composition (mg/g) of groups. Values are mean ± SE. 'Presence': the limpet partially covered by the coralline algae, 'Absence': the limpet without the coralline algae on the shell.

Fatty Acid	Trivial Name	Presence N = 20	Absence N = 20	p-Value
C14:0	Myristic acid (tetradecanoic acid)	0.192 (±0.007)	0.178 (±0.006)	0.145
C16:0	Palmitic acid (hexadecanoic acid)	3.463 (±0.061)	3.228 (±0.038)	0.002 *
C17:0	Margaric acid (heptadecanoic acid)	0.248 (±0.010)	0.179 (±0.004)	<0.001 *
C18:0	Stearic acid (octadecanoic acid)	0.952 (±0.020)	0.868 (±0.011)	0.001 *
Σ SFA [1]	-	4.855 (±0.090)	4.452 (±0.050)	<0.001 *
C18:1	Oleic acid	0.531 (±0.021)	0.536 (±0.016)	0.84
C20:1	Eicosenoic acid	0.992 (±0.027)	1.065 (±0.025)	0.052
Σ MUFA [2]	-	1.523 (±0.044)	1.602 (±0.031)	0.153

Table 1. *Cont.*

Fatty Acid	Trivial Name	Presence $N = 20$	Absence $N = 20$	p-Value
C18:2	Linoleic acid	0.062 (±0.006)	0.093 (±0.006)	0.001 *
C18:3	Alpha-linolenic acid (ALA)	0.131 (±0.015)	0.222 (±0.014)	<0.001 *
C20:2	Eicosadienoic acid	0.928 (±0.039)	0.951 (±0.026)	0.613
C20:3	Eicosatrienoic acid (ETE)	0.817 (±0.028)	0.842 (±0.019)	0.478
C20:4	Arachidonic acid (AA)	2.240 (±0.089)	2.451 (±0.061)	0.058
C20:5	Eicosapentaenoic acid (EPA, Timnodonic acid)	4.039 (±0.174)	3.948 (±0.091)	0.648
Σ PUFA [3]	-	8.217 (±0.281)	8.507 (±0.091)	0.337
Σ n–3 [4]	-	1.662 (±0.072)	1.671 (±0.041)	0.912
Σ n–6 [5]	-	1.077 (±0.044)	1.165 (±0.031)	0.089
n–3:n–6	-	1.565 (±0.063)	1.457 (±0.056)	0.212

[1] SFA, Saturated Fatty Acid; [2] MUFA, Monounsaturated Fatty Acid; [3] PUFA, Polyunsaturated Fatty Acid; [4] n–3, n–3 Fatty acid; [5] n–6, n–6 Fatty acid; * Significant p values with sequential Bonferroni correction ($p < 0.05$).

3.3. Scanning Electron Microscope (SEM) Analysis

SEM observation showed that the algae had no direct effect on the shell. No damage was observed to the surface shells from which the algae were removed (Figure 2c). However, the surfaces of the shells without algae were damaged, with irregular holes observed (Figure 2d).

3.4. Environmental Factors and Limpet

A significantly positive correlation was found between the coverage (%) of coralline algae on the limpet and the temperature and pH of the habitat seawater (Figure 4 and Table 2).

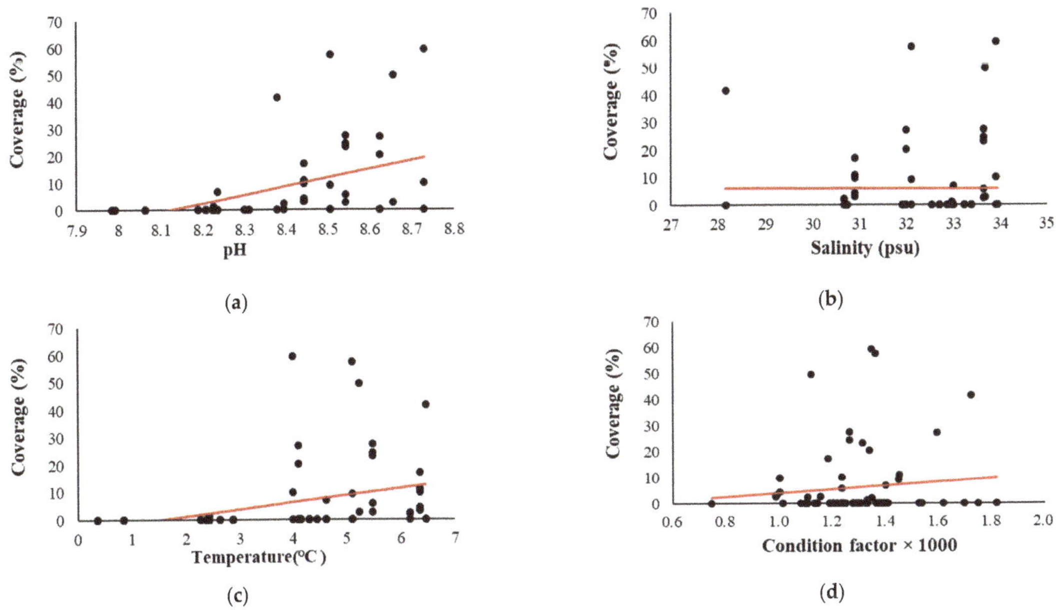

Figure 4. Relationship between (**a**) pH, (**b**) temperature, (**c**) salinity, and (**d**) CF and the coverage of coralline algae on the limpet shell. The red line indicates the trend line.

Table 2. Spearman's rank correlation analysis examines the relationship between pH, temperature, salinity, and the CF of the limpets and the coverage of coralline algae on the limpet shell. Description of symbols: r—Spearman's Rank Correlation Coefficient; *N*—number of data; *p*—the *p*-value of the correlation.

	pH	Temperature	Salinity	Condition Factor × 1000
Coverage	$y = 31.876x - 258.95$ $r = 0.664, N = 69, p < 0.001$	$y = 2.5621x - 4.0368$ $r = 0.539, N = 69, p < 0.001$	$y = -0.0349x + 7.1865$ $r = 0.036, N = 69, p = 0.772$	$y = 7.1391x - 3.1859$ $r = -0.055, N = 69, p = 0.654$

4. Discussion

In this study, we found that Antarctic limpets with coralline algae epibionts had higher mortality and greater shell weight than limpets without coralline algae and identified variations in the ratio of fatty acids between the two groups. However, the CF did not differ significantly between the two groups. These results suggest that coralline algae act as a parasite under normal conditions, with their presence negatively affecting behavior and metabolic activity by increasing the shell weight of the host and by decreasing drag speed [32], eventually jeopardizing their survival [33]. Our study showed that the shell weight of the limpet was higher in the group with coralline algae than in the group without coralline algae. An increase in shell weight can increase the metabolic burden of the host organism, eventually leading to energy exhaustion [32].

Although the CF could be a sensitive indicator for other organisms that generally inhabit the intertidal zone, the CF of *N. concinna* did not respond sensitively to pH [25] and temperature [34] in previous studies. Therefore, it appears that epibionts do not affect the CF of the limpet.

The fatty acid composition, which can be an indicator of stress (temperature, salinity, pollutants, etc.) in living organisms, was found to vary significantly depending on the presence or absence of epibionts [35,36]. In the group with coralline algae, the C16:0, C17:0, C18:0, and total SFA values were higher than in the group without coralline algae. The PUFA values of C18:2n6c, C18:3n3, and PUFA: SFA were lower in the group with coralline algae. It is well known that changes in the fatty acid composition are one of the cell-unit protection strategies that protect against environmental changes [37]. Changes in fatty acids were also seen in the military turban sea snail (*Turbo militaris*) [38] with increasing temperature and in the blue mussel (*Mytilus edulis* L.) [39] with increasing salinity.

Changes in SFA and PUFA are characteristics that regulate membrane structure and features in organisms to resist environmental stress, respectively [37,40–42]. This study has shown that the presence of an epibiont can change the fatty acid composition of the host and may also influence the cell functioning.

SEM analysis showed that the degree of damage to the surface of the limpet shell was different depending on the presence or absence of coralline algae. Limpet shells can be damaged by environmental factors such as waves and glaciers, as well as biological factors such as endolithic algae [43]. Endolithic algae are very important microborers for organisms with carbonate exoskeletons [44–46]. However, our study could not confirm whether the coralline algae directly affected the shell of the limpet.

Interactions between epibionts and hosts can be altered by environmental changes [47,48]. For instance, while branchiobdellids, which are epibionts of crayfish, act as commensals in clean water, their relationship changes to mutualism under a fouling environment [47]. The limpets and coralline algae are sensitive to ocean acidification in Antarctica because calcium carbonate ($CaCO_3$) makes up the main component of their shell [19,49]. When the temperature and pH of the seawater fluctuate due to climate change, the epibiotic relationship between limpet and coralline algae will also be affected. In our study, the area of coralline algae on the limpets showed an opposite trend with decreasing pH and increasing temperature. Despite this, studies have shown that endolithic algae can thrive when seawater pH decreases and water temperature increases [50–52]. In addition, it was suggested that increasing the biomass of microborers can partially improve the dissolution rate (%) of coral exoskeletons [53]. Therefore, if endolithic algae are more likely to proliferate on limpets because of climate change, coralline algae can protect the shell because they adhere to the outermost surface of the limpet shell. This can lead to a mutually beneficial symbiosis as the benefits received by the limpets in the existing parasitic relationship will be substantially increased.

5. Conclusions

This study evaluated the interactions between the Antarctic limpet, representative intertidal macrofauna living in the Antarctic intertidal zone, and coralline algae fouled on the shells. Negative effects of this relationship were that the mortality and shell weight of the limpets were increased, and the distribution of the fatty acids was altered. As a positive effect, the algae physically protected the limpet shell from external stressors (scratches, penetration, etc.). We provide the first evidence that the relationship between the Antarctic limpet (*N. concinna*) and coralline algae is parasitic. However, future climate change may alter this relationship. Further studies on the effects of climate change on the interactions between these two species would help understand how symbiotic relationships can develop between species.

Author Contributions: B.C., H.B. and T.K. conceived and designed the experiments; B.C. and H.B. performed the experiments; B.C. and H.B. analyzed the data; B.C. and H.B. contributed reagents/materials/analysis tools; B.C. and H.B. wrote the paper; T.K. conducted writing-review and editing; T.K. supervised the research and in charge funding acquisition. All authors have read and agreed to the published version of the manuscript.

Funding: This work was supported by Korea Polar Research Institute under grant PE22110 and Inha University research grant.

Institutional Review Board Statement: Not applicable.

Informed Consent Statement: Not applicable.

Data Availability Statement: The data that support the findings of this study are available from the corresponding author upon reasonable request.

Acknowledgments: We thank Ji Hee Kim of the Korea Polar Research Institute for identifying the coralline algae species.

Conflicts of Interest: The authors declare no conflict of interest.

References

1. Wahl, M. Marine Epibiosis. I. Fouling and Antifouling: Some Basic Aspects. *Mar. Ecol. Prog. Ser.* **1989**, *58*, 175–189. [CrossRef]
2. Wahl, M. (Ed.) Epibiosis: Ecology, Effects and Defenses. In *Marine Hard Bottom Communities: Patterns, Dynamics, Diversity, and Change*; Springer Series: Ecological Studies, 206; Springer: Berlin/Heidelberg, Germany, 2009; pp. 61–72. ISBN 978-3-540-92703-7.
3. Bach, C.; Hazlett, B.; Rittschof, D. Sex-Specific Differences and the Role of Predation in the Interaction between the Hermit Crab, *Pagurus Longicarpus*, and Its Epibiont, *Hydractinia Symbiolongicarpus*. *J. Exp. Mar. Biol. Ecol.* **2006**, *333*, 181–189. [CrossRef]
4. Buschbaum, C.; Buschbaum, G.; Schrey, I.; Thieltges, D. Shell-Boring Polychaetes Affect Gastropod Shell Strength and Crab Predation. *Mar. Ecol.-Prog. Ser.* **2007**, *329*, 123–130. [CrossRef]
5. Enderlein, P.; Moorthi, S.; Rohrscheidt, H.; Wahl, M. Optimal Foraging versus Shared Doom Effects: Interactive Influence of Mussel Size and Epibiosis on Predator Preference. *J. Exp. Mar. Biol. Ecol.* **2003**, *292*, 231–242. [CrossRef]
6. Manning, L.M.; Lindquist, N. Helpful Habitant or Pernicious Passenger: Interactions between an Infaunal Bivalve, an Epifaunal Hydroid and Three Potential Predators. *Oecologia* **2003**, *134*, 415–422. [CrossRef]
7. Marin, A.; López Belluga, M.D. Sponge Coating Decreases Predation On The Bivalve Arca Noae. *J. Molluscan Stud.* **2005**, *71*, 1–6. [CrossRef]
8. Wahl, M.; Hay, M.E. Associational Resistance and Shared Doom: Effects of Epibiosis on Herbivory. *Oecologia* **1995**, *102*, 329–340. [CrossRef]
9. Cerrano, C.; Puce, S.; Chiantore, M.; Bavestrello, G.; Cattaneo-Vietti, R. The Influence of the Epizoic Hydroid *Hydractinia Angusta* on the Recruitment of the Antarctic Scallop *Adamussium Colbecki*. *Polar Biol.* **2001**, *24*, 577–581. [CrossRef]
10. Cerrano, C.; Bertolino, M.; Valisano, L.; Bavestrello, G.; Calcinai, B. Epibiotic Demosponges on the Antarctic Scallop *Adamussium Colbecki* (Smith, 1902) and the Cidaroid Urchins *Ctenocidaris Perrieri* Koehler, 1912 in the Nearshore Habitats of the Victoria Land, Ross Sea, Antarctica. *Polar Biol.* **2009**, *32*, 1067–1076. [CrossRef]

11. Gutt, J.; Schickan, T. Epibiotic Relationships in the Antarctic Benthos. *Antarct. Sci.* **1998**, *10*, 398–405. [CrossRef]
12. Jones, E.M.; Fenton, M.; Meredith, M.P.; Clargo, N.M.; Ossebaar, S.; Ducklow, H.W.; Venables, H.J.; de Baar, H.J.W. Ocean Acidification and Calcium Carbonate Saturation States in the Coastal Zone of the West Antarctic Peninsula. *Deep Sea Res. Part II Top. Stud. Oceanogr.* **2017**, *139*, 181–194. [CrossRef]
13. Montes, M.; Doney, S.; Ducklow, H.; Fraser, W.; Martinson, D.; Stammerjohn, S.; Schofield, O. Recent Changes in Phytoplankton Communities Associated with Rapid Regional Climate Change Along the Western Antarctic Peninsula. *Science* **2009**, *323*, 1470–1473. [CrossRef] [PubMed]
14. Kim, B.-M.; Lee, Y.; Hwang, J.-Y.; Kim, Y.-K.; Kim, T.W.; Kim, I.-N.; Kang, S.; Kim, J.-H.; Rhee, J.-S. Physiological and Molecular Responses of the Antarctic Harpacticoid Copepod *Tigriopus Kingsejongensis* to Salinity Fluctuations—A Multigenerational Study. *Environ. Res.* **2022**, *204*, 112075. [CrossRef] [PubMed]
15. Servetto, N.; de Aranzamendi, M.C.; Bettencourt, R.; Held, C.; Abele, D.; Movilla, J.; González, G.; Bustos, D.M.; Sahade, R. Molecular Mechanisms Underlying Responses of the Antarctic Coral *Malacobelemnon Daytoni* to Ocean Acidification. *Mar. Environ. Res.* **2021**, *170*, 105430. [CrossRef] [PubMed]
16. de Aranzamendi, M.C.; Servetto, N.; Movilla, J.; Bettencourt, R.; Sahade, R. Ocean Acidification Effects on the Stress Response in a Calcifying Antarctic Coastal Organism: The Case of *Nacella Concinna* Ecotypes. *Mar. Pollut. Bull.* **2021**, *166*, 112218. [CrossRef] [PubMed]
17. Park, S.; Ahn, I.-Y.; Sin, E.; Shim, J.; Kim, T. Ocean Freshening and Acidification Differentially Influence Mortality and Behavior of the Antarctic Amphipod *Gondogeneia Antarctica*. *Mar. Environ. Res.* **2020**, *154*, 104847. [CrossRef]
18. Picken, G.B. The Distribution, Growth, and Reproduction of the Antarctic Limpet *Nacella (Patinigera) Concinna* (Strebel, 1908). *J. Exp. Mar. Biol. Ecol.* **1980**, *42*, 71–85. [CrossRef]
19. McClintock, J.B.; Angus, R.A.; Mcdonald, M.R.; Amsler, C.D.; Catledge, S.A.; Vohra, Y.K. Rapid Dissolution of Shells of Weakly Calcified Antarctic Benthic Macroorganisms Indicates High Vulnerability to Ocean Acidification. *Antarct. Sci.* **2009**, *21*, 449–456. [CrossRef]
20. Nolan, C.P. Size, Shape and Shell Morphology in the Antarctic Limpet *Nacella concinna* at Signy Island, South Orkney Islands. *J. Molluscan Stud.* **1991**, *57*, 225–238. [CrossRef]
21. Wulff, A.; Iken, K.; Quartino, M.; Al-Handal, A.; Wiencke, C.; Clayton, M. Biodiversity, Biogeography and Zonation of Marine Benthic Micro- and Macroalgae in the Arctic and Antarctic. *Bot. Mar.* **2009**, *56*, 491–507. [CrossRef]
22. Schoenrock, K.M.; Schram, J.B.; Amsler, C.D.; McClintock, J.B.; Angus, R.A.; Vohra, Y.K. Climate Change Confers a Potential Advantage to Fleshy Antarctic Crustose Macroalgae over Calcified Species. *J. Exp. Mar. Biol. Ecol.* **2016**, *474*, 58–66. [CrossRef]
23. Wiencke, C. *Biology of Polar Benthic Algae*; Walter de Gruyter: Berlin, Germany, 2011; ISBN 978-3-11-022970-7.
24. Adey, W.H.; Hernandez-Kantun, J.J.; Johnson, G.; Gabrielson, P.W. DNA Sequencing, Anatomy, and Calcification Patterns Support a Monophyletic, Subarctic, Carbonate Reef-Forming Clathromorphum (Hapalidiaceae, Corallinales, Rhodophyta). *J. Phycol.* **2015**, *51*, 189–203. [CrossRef] [PubMed]
25. Sin, E.; Ahn, I.-Y.; Park, S.; Kim, T. Effects of Low PH and Low Salinity Induced by Meltwater Inflow on the Behavior and Physical Condition of the Antarctic Limpet, *Nacella Concinna*. *J. Mar. Sci. Eng.* **2020**, *8*, 822. [CrossRef]
26. Saad, A.E.-H.A. Age, Growth and Morphometry of the Limpet *Cellana Eucosmia* (Mollusca: Gastropoda) from the Gulf of Suez. *Indian J. Geo-Mar. Sci.* **1997**, *262*, 169–172.
27. González, P.M.; Puntarulo, S.; Malanga, G. Natural Adaptation to the Environmental Conditions Affects the Oxidation-Dependent Processes in Limpets. *Int. J. Environ. Health* **2017**, *8*, 302–315. [CrossRef]
28. Garces, R.; Mancha, M. One-Step Lipid Extraction and Fatty Acid Methyl Esters Preparation from Fresh Plant Tissues. *Anal. Biochem.* **1993**, *211*, 139–143. [CrossRef]
29. Tan, K.; Zhang, H.; Ma, H.; Li, S.; Zheng, H. Effects of Tidal Zones and Seasons on Nutritional Properties of Commercially Importance Gastropods. *Estuar. Coast. Shelf Sci.* **2021**, *254*, 107289. [CrossRef]
30. Valles-Regino, R.; Tate, R.; Kelaher, B.; Savins, D.; Dowell, A.; Benkendorff, K. Ocean Warming and CO2-Induced Acidification Impact the Lipid Content of a Marine Predatory Gastropod. *Mar. Drugs* **2015**, *13*, 6019–6037. [CrossRef]
31. Rice, W.R. Analyzing Tables of Statistical Tests. *Evolution* **1989**, *43*, 223–225. [CrossRef]
32. Wahl, M. Increased Drag Reduces Growth of Snails: Comparison of Flume and in Situ Experiments. *Mar. Ecol. Prog. Ser.* **1997**, *151*, 291–293. [CrossRef]
33. Thieltges, D.W. Impact of an Invader: Epizootic American Slipper Limpet *Crepidula Fornicata* Reduces Survival and Growth in European Mussels. *Mar. Ecol. Prog. Ser.* **2005**, *286*, 13–19. [CrossRef]
34. Obermüller, B.E.; Morley, S.A.; Clark, M.S.; Barnes, D.K.A.; Peck, L.S. Antarctic Intertidal Limpet Ecophysiology: A Winter–Summer Comparison. *J. Exp. Mar. Biol. Ecol.* **2011**, *403*, 39–45. [CrossRef]
35. Ericson, J.A.; Hellessey, N.; Kawaguchi, S.; Nichols, P.D.; Nicol, S.; Hoem, N.; Virtue, P. Near-Future Ocean Acidification Does Not Alter the Lipid Content and Fatty Acid Composition of Adult Antarctic Krill. *Sci. Rep.* **2019**, *9*, 12375. [CrossRef] [PubMed]
36. Puccinelli, E.; McQuaid, C.D. Commensalism, Antagonism or Mutualism? Effects of Epibiosis on the Trophic Relationships of Mussels and Epibiotic Barnacles. *J. Exp. Mar. Biol. Ecol.* **2021**, *540*, 151549. [CrossRef]
37. Fokina, N.N.; Ruokolainen, T.R.; Nemova, N.N.; Bakhmet, I.N. Changes of Blue Mussels *Mytilus Edulis* L. Lipid Composition under Cadmium and Copper Toxic Effect. *Biol. Trace Elem. Res.* **2013**, *154*, 217–225. [CrossRef] [PubMed]

38. Ab Lah, R.; Kelaher, B.P.; Bucher, D.; Benkendorff, K. Ocean Warming and Acidification Affect the Nutritional Quality of the Commercially-Harvested Turbinid Snail Turbo Militaris. *Mar. Environ. Res.* **2018**, *141*, 100–108. [CrossRef] [PubMed]
39. Nemova, N.N.; Fokina, N.N.; Nefedova, Z.A.; Ruokolainen, T.R.; Bakhmet, I.N. Modifications of Gill Lipid Composition in Littoral and Cultured Blue Mussels *Mytilus Edulis* L. under the Influence of Ambient Salinity. *Polar Rec.* **2013**, *49*, 272–277. [CrossRef]
40. Babaran, D.; Koprivnikar, J.; Parzanini, C.; Arts, M.T. Parasites and Their Freshwater Snail Hosts Maintain Their Nutritional Value for Essential Fatty Acids despite Altered Algal Diets. *Oecologia* **2021**, *196*, 553–564. [CrossRef]
41. Tocher, D.R. Metabolism and Functions of Lipids and Fatty Acids in Teleost Fish. *Rev. Fish. Sci.* **2003**, *11*, 107–184. [CrossRef]
42. De Carvalho, C.C.C.R.; Caramujo, M.J. The Various Roles of Fatty Acids. *Molecules* **2018**, *23*, 2583. [CrossRef]
43. Cadee, G. Shell Damage and Shell Repair in the Antarctic Limpet *Nacella Concinna* from King George Island. *J. Sea Res.* **1999**, *41*, 149–161. [CrossRef]
44. Golubic, S.; Friedmann, E.I.; Schneider, J. The Lithobiontic Ecological Niche, with Special Reference to Microorganisms. *J. Sediment. Res.* **1981**, *51*, 475–478. [CrossRef]
45. Le Campion-Alsumard, T.; Golubic, S.; Hutchings, P. Microbial Endoliths in Skeletons of Live and Dead Corals: Porites Lobata (Moorea, French Polynesia). *Mar. Ecol. Prog. Ser.* **1995**, *117*, 149–157. [CrossRef]
46. Verbruggen, H.; Tribollet, A. Boring Algae. *Curr. Biol.* **2011**, *21*, R876–R877. [CrossRef]
47. Lee, J.H.; Kim, T.W.; Choe, J.C. Commensalism or Mutualism: Conditional Outcomes in a Branchiobdellid–Crayfish Symbiosis. *Oecologia* **2009**, *159*, 217–224. [CrossRef]
48. Kagawa, O.; Uchida, S.; Yamazaki, D.; Osawa, Y.; Ito, S.; Chiba, S. Citizen Science via Social Media Revealed Conditions of Symbiosis between a Marine Gastropod and an Epibiotic Alga. *Sci. Rep.* **2020**, *10*, 19647. [CrossRef]
49. Guinotte, J.M.; Fabry, V.J. Ocean Acidification and Its Potential Effects on Marine Ecosystems. *Ann. N. Y. Acad. Sci.* **2008**, *1134*, 320–342. [CrossRef]
50. Diaz-Pulido, G.; Anthony, K.R.N.; Kline, D.I.; Dove, S.; Hoegh-Guldberg, O. Interactions Between Ocean Acidification and Warming on the Mortality and Dissolution of Coralline Algae1. *J. Phycol.* **2012**, *48*, 32–39. [CrossRef]
51. Fine, M.; Loya, Y. Endolithic Algae: An Alternative Source of Photoassimilates during Coral Bleaching. *Proc. R. Soc. Lond. B Biol. Sci.* **2002**, *269*, 1205–1210. [CrossRef]
52. Tribollet, A.; Godinot, C.; Atkinson, M.; Langdon, C. Effects of Elevated PCO2 on Dissolution of Coral Carbonates by Microbial Euendoliths. *Glob. Biogeochem. Cycles* **2009**, *23*, GB3008. [CrossRef]
53. Reyes-Nivia, C.; Diaz-Pulido, G.; Kline, D.; Guldberg, O.-H.; Dove, S. Ocean Acidification and Warming Scenarios Increase Microbioerosion of Coral Skeletons. *Glob. Change Biol.* **2013**, *19*, 1919–1929. [CrossRef] [PubMed]

Journal of *Marine Science and Engineering*

MDPI

Article

Identification and Phylogenetic Analysis of Chitin Synthase Genes from the Deep-Sea Polychaete *Branchipolynoe onnuriensis* Genome

Hyeongwoo Choi [1], Sang Lyeol Kim [2,3], Man-Ki Jeong [4], Ok Hwan Yu [3] and Seongil Eyun [1,*]

1 Department of Life Science, Chung-Ang University, Seoul 06974, Korea; creo9447@cau.ac.kr
2 Ocean Science and Technology School, Korea Maritime University, Busan 49111, Korea; boyis20c@kiost.ac.kr
3 Marine Ecosystem and Biological Research Centre, Korea Institute of Ocean Science and Technology, Busan 49111, Korea; ohyu@kiost.ac.kr
4 Department of Smart Fisheries Resources Management, Chonnam National University, Yeosu 59626, Korea; jmgdeux@chonnam.ac.kr
* Correspondence: eyun@cau.ac.kr

Abstract: Chitin, one of the most abundant biopolymers in nature, is a crucial material that provides sufficient rigidity to the exoskeleton. In addition, chitin is a valuable substance in both the medical and industrial fields. The synthesis of chitin is catalyzed by chitin synthase (*CHS*) enzymes. Although the chitin synthesis pathway is highly conserved from fungi to invertebrates, *CHSs* have mostly only been investigated in insects and crustaceans. Especially, little is known about annelids from hydrothermal vents. To understand chitin synthesis from the evolutionary view in a deep-sea environment, we first generated the whole-genome sequencing of the parasitic polychaete *Branchipolynoe onnuriensis*. We identified seven putative *CHS* genes (*BonCHS1-BonCHS7*) by domain searches and phylogenetic analyses. This study showed that most crustaceans have only a single copy or two gene copies, whereas at least two independent gene duplication events occur in *B. onnuriensis*. This is the first study of *CHS* obtained from a parasitic species inhabiting a hydrothermal vent and will provide insight into various organisms' adaptation to the deep-sea hosts.

Keywords: chitin synthase; polychaete; *Branchipolynoe*; host–parasite interaction

1. Introduction

Chitin, a linear polymer of β-(1,4)-*N*-acetyl-D-glucosamine (GlcNac), is the second most abundant biopolymer in nature, followed by cellulose, with more than 100 billion tons synthesized annually [1–5]. It is found in various organisms, ranging from fungi to various invertebrates, and provides them with sufficient rigidity to support their shape and structure [1]. In arthropods, chitin plays a crucial role in forming new cuticles during molting and is a component of the intestine peritrophic matrix in insects, which supports digestion [1,6,7]. In nematodes, chitin components are found in the eggshell and pharynx [8]. Furthermore, in Lophotrochozoa, chitin forms the radula and shell in mollusks [9–11], beak in cephalopods [12], and chaetae in annelids [13]. Due to its diverse function, chitin is attracting attention as a raw material for various fields, such as the pharmaceutical and biotechnological industries [14].

Chitin is polymerized by an enzyme called chitin synthase (*CHS*, chitin 4-β-*N*-acetylglucosaminyltransferase; EC 2.4.1.16), which is generally characterized by three functional domains: A, B, and C [15]. Domain A, composed of several transmembrane helixes, is located at the N-terminal, and this domain sequence may vary between species. Domain B (chitin_synth_2), the catalytic core that contains two highly conserved motifs ("EDR" and "QRRRW"), is in the middle of the gene. Domain C is located at the C-terminal, with approximately seven transmembrane helices, and has the conserved motif "WGTRE" [1].

Generally, insects have two *CHS* genes (*CHS1* and *CHS2*). *CHS1* is responsible for cuticle formation in the epidermis, while *CHS2* is involved in chitin synthesis in the peritrophic membrane of the intestine [1,6]. A *CHS* gene knockdown study in the crustacean *Lepeophtheirus salmonis* showed the formation of an abnormal appendage, which eventually led to death, suggesting the multifunctional role of *CHS* [16]. However, interestingly, compared to the ecdysozoans, which have only a single or two gene copies located in the same chromosome, numerous *CHS* genes have been identified in lophotrochozoans [1]. For example, 31 *CHS* genes were identified in the brachiopod *Lingula anatina* [17]. In addition, four and five *CHS* genes were identified in the shallow-sea polychaetes *Capitella teleta* and *Dimorphilus gyrociliatus*, respectively, whereas 19 and 12 *CHS* genes were significantly expanded in the deep-sea polychaetes *Paraescarpia echinospica* and *Lamellibrachia luymesi* [18]. These findings suggest that the *CHS* gene duplication event occurs lineage-specifically. However, *CHSs* have mostly been explored only in arthropods; so far, little is known about *CHSs* in annelids. Studies are gradually investigating the poorly explored realm of the lophotrochozoan *CHS*, but data on the evolutionary process of *CHS* gene expansion are still lacking. In addition, no phylogenetic analysis, including deep-sea parasitic polychaetes, has been reported.

In this study, we performed whole-genome sequencing (WGS) of parasitic polychaete *Branchipolynoe onnuriensis* collected from bivalves living in a hydrothermal vent [19] and identified seven *CHS* genes (*BonCHS1–BonCHS7*) belonging to the glycosyltransferase 2 (GT2) family. This is the first study of *CHSs* from deep-sea parasitic polychaetes. We also analyzed the relationship of *B. onnuriensis CHS* genes with those from lophotrochozoans. In addition, we expanded on the classification of lophotrochozoan *CHS* gene groups, in order to obtain information about their *CHS* gene family expansion and categorized them into five different subgroups. Our results will provide important information for those who study the chitin synthesis mechanism in deep-sea parasitic polychaetes in the future.

2. Materials and Methods

2.1. Sample Collection and Next-Generation Sequencing

An individual parasitic polychaete *Branchipolynoe onnuriensis* was separated from its host *Gigantidas vrijenhoeki* (class Bivalvia) using a video-guided hydraulic grab (Oktopus, Germany) around the Onnuri Vent Field (OVF, 11°14′55.92″ S, 66°15′15.10″ E; depth of 2014.5 m) during a Korea Institute of Ocean Science and Technology (KIOST) expedition along the Central Indian Ridge (CIR) in 2019 [20]. Immediately after being collected, the sample was stored in 95% ethanol at −20 °C, until DNA extraction in the laboratory. Genomic DNA was extracted using the QIAGEN Blood & Cell Culture DNA Mini Kit (QIAGEN, Hilden, Germany), according to the manufacturer's instructions. A paired-end library was constructed using the TruSeq DNA Nano 550 bp kit (Illumina, Inc., San Diego, CA, USA), with an insert size of 550 bp, and 150 bp sequencing was performed using the Novaseq6000 platform (Illumina).

2.2. Data Filtering and De Novo Genome Assembly

Adaptor sequences and low-quality reads that were lower than the mean quality score of 20 were removed. In addition, reads shorter than 120 bp or with unknown bases (n) were filtered using Trim Galore (ver. 0.6.6) [21]. The cleaned reads were obtained with the following parameters: `-quality 20 -length 120 -max_n 0`.

After quality control, de novo assembly, using 21-, 31-, and 51-mers to build an initial *de Bruijn* graph, was performed with SPAdes (ver. 3.14.0) [22]. Finally, quality assessment software for genome assembly, QUAST (ver. 5.0.2) were utilized to obtain diverse metrics, such as the number of contigs, a large contig, the total length, N50, or L50, without a reference genome [23].

2.3. Gene Prediction and Identification of the Chitin Synthase Gene

The genome structure of *B. onnuriensis* was annotated using ab initio gene prediction with Augustus (ver. 3.4.0) using a generalized hidden Markov model [24].

To extract the putative CHS sequences, we combined the basic local alignment search tool (BLAST) searches using National Center for Biotechnology Information (NCBI) and domain predictions. According to Zakrzewski et al. (2014) [7], since lophotrochozoans have four subgroups of *CHS* genes in type 2, we assumed that there would be at least one gene in each group (A, B, C, and D). First, we mined five *CHS* genes corresponding to each group from the same polychaete species, three *CHS* genes from *Owenia fusiformis* (group A, accession no. AHX26704.1; group D, accession no. AHX26707.1; type 1, accession no. AHX26703.1), and two *CHS* genes from *Sabellaria alveolata* (group B, accession no. AHX26717.1; group C, accession no. AHX26711.1) from NCBI and used them as queries to search homologous genes in our sample. In addition, we performed BLAST searches against the customized database with an *E*-value cut-off of $< 1 \times 10^{-50}$ and a length of > 300 amino acids (aa) [25]. Next, domain searches of each obtained gene were carried out using the simple modular architecture research tool (SMART) [26]. We identified seven putative genes and named them *BonCHS1–BonCHS7*. To confirm the putative BonCHSs, we performed BLAST searches against the Carbohydrate-Active enZymes Database (CAZy; https://bcb.unl.edu/dbCAN2/download/CAZyDB.09242021.fa, accessed on 1 June 2022) which contains enzymes that synthesize or break down carbohydrates and glycoconjugates, with an *E*-value cut-off of $< 1 \times 10^{-100}$ [27]. In order to obtain comparable E-values, the database size of 1.58×10^{11} (using the "`-dbsize`" option) was set to be equivalent to the size of the non-redundant (NR) protein database at NCBI.

2.4. Phylogenetic Analysis of Chitin Synthase

We conducted two phylogenetic analyses, i.e., the lophotrochozoan and metazoan trees. In the lophotrochozoan phylogenetic tree, 52 protein sequences, including seven putative *BonCHS1-BonCHS7*, were retrieved from 20 species (Table 1). In the metazoan *CHS* gene tree, 74 protein sequences were retrieved from 33 species (Table 1). Multiple sequence alignments were performed with MAFFT (ver. 7.475) [28]. We also used IQ-TREE (ver. 2.2.0) to select the best substitution model via Bayesian information criterion (BIC) [29]. The best evolutionary model of LG + I + G4 and LG + F + I + G4 was selected to construct the maximum likelihood (ML) for metazoan and lophotrochozoan data sets using RAxML-NG (ver. 0.9.0) [30]. In addition, each branch was supported by 1000 bootstrap replicates in the ML tree. The Bayesian tree was constructed using MrBayes (ver. 3.2.4), using the LG + I + G4 and LG + F + I + G4 models for metazoan and lophotrochozoan datasets. A total of four chains (three heated and one cold chain), for sampling all 5×10^2 generations, were carried out in two independent analyses. We performed 1×10^6 generations of MCMC analysis, and the first 25% trees as burn-in and incorporated with the ML tree [31]. Finally, each metazoan and lophotrochozoan phylogenetic tree was visualized using FigTree (ver. 1.4.4).

Table 1. Gene list of *chitin synthases* used for phylogenetic analysis.

Taxon (Phylum: Class)	Species	Gene Name	Type	Accession No.
Annelida *: Polychaeta	*Owenia fusiformis*	*OfuCHS1*	Type 2, group A	AHX26704.1
		OfuCHS2	Type 2, group C	AHX26705.1
		OfuCHS3	Type 2, group C	AHX26706.1
		OfuCHS4	Type 2, group D	AHX26707.1
		OfuCHS5	Type 2, group D	AHX26713.1
		OfuCHS6	Type 1	AHX26703.1
	Sabellaria alveolata	*SalCHS1*	Type 2, group B	AHX26717.1
		SalCHS2	Type 2, group C	AHX26711.1
		SalCHS3	Type 1	AHX26710.1
	Dimorphilus gyrociliatus	*DgyCHS1*	Type 2, group A	CAD5118528.1
		DgyCHS2	Type 2, group B	CAD5114651.1

Table 1. *Cont.*

Taxon (Phylum: Class)	Species	Gene Name	Type	Accession No.
	Platynereis dumerilii	*PduCHS1*	Type 2, group B	AHX26708.1
		PduCHS2	Type 2, group B	AHX26709.1
		PduCHS3	Type 2, group C	AHX26716.1
	Capitella teleta	*CteCHS1*	Type 2, group A	ELU08572.1
		CteCHS2	Type 2, group C	ELT98539.1
		CteCHS3	Type 2, group D	ELT92724.1
		CteCHS4	Type 2, group D	ELT92107.1
Annelida *: Annelida *incertae sedis*	*Myzostoma cirriferum*	*MciCHS1*	Type 2, group B	AHX26714.1
Mollusca *: Bivalvia	*Lottia gigantea*	*LgiCHS1*	Type 2, group A	XP_009061726.1
		LgiCHS2	Type 2, group A	XP_009061725.1
		LgiCHS3	Type 2, group A	XP_009061724.1
		LgiCHS4	Type 2, group B	XP_009063632.1
		LgiCHS5	Type 2, group C	XP_009047936.1
		LgiCHS6	Type 2, group D	XP_009066852.1
		LgiCHS7	Type 2, group D	XP_009066854.1
		LgiCHS8	Type 2, group D	XP_009051436.1
		LgiCHS9	Type 2, group D	XP_009051165.1
	Mytilus edulis	*MedCHS1*	Type 2, group A	CAG2205753.1
	Tegillarca granosa	*TgrCHS1*	Type 2, group A	AON76719.1
	Atrina rigida	*AriCHS1*	Type 2, group A	AAY86556.1
	Pinctada fucata	*PfuCHS1*	Type 2, group A	BAF73720.1
Mollusca *: Gastropoda	*Elysia marginata*	*EmaCHS1*	Type 2, group A	GFS24687.1
		EmaCHS2	Type 2, group C	GFR89942.1
		EmaCHS3	Type 2, group D	GFR83755.1
		EmaCHS4	Type 2, group D	GFR70591.1
		EmaCHS5	Type 1	GFS00558.1
		EmaCHS6	Type 1	GFR82903.1
	Crassostrea gigas	*CgiCHS1*	Type 2, group A	XP_034323514.1
Mollusca *: Polyplacophora	*Leptochiton asellus*	*LasCHS1*	Type 2, group A	AHX26699.1
		LasCHS2	Type 2, group C	AHX26700.1
Arthropoda: Insecta	*Helicoverpa zea*	*HzeCHS1*	Type 2, group 1	ADX66429.1
		HzeCHS2	Type 2, group 2	ADX66427.1
	Ostrinia furnacalis	*OfurCHS1*	Type 2, group 1	ACF53745.1
		OfurCHS2	Type 2, group 2	ABB97082.1
	Manduca sexta	*MseCHS1*	Type 2, group 1	AAL38051.2
		MseCHS2	Type 2, group 2	AAX20091.1
Arthropoda: Copepoda	*Lepidopterous salmonis*	*LsaCHS1*	Type 2, group 1	AYN59157.1
		LsaCHS2	Type 2, group 2	AYN59158.1
Chordata: Actinopterygii	*Danio rerio*	*DreCHS1*	Type 2 deuterostome	AJW72838.1
	Esox lucius	*EluCHS1*	Type 2 deuterostome	XP_010887243.2
	Megalops cyprinoides	*McyCHS1*	Type 2 deuterostome	XP_036403039.1
Chordata: Ascidiacea	*Ciona robusta*	*CroCHS1*	Type 2 deuterostome	BBB15954.1
Chordata: Amphibia	*Xenopus laevis*	*XlaCHS1*	Type 2 deuterostome	XP_018120159.2
Choanoflagellatea	*Salpingoeca rosetta*	*SroCHS1*	Type 1	EGD80959.1
	Monosiga brevicollis	*MbrCHS1*	Type 1	XP_001743227.1

Table 1. *Cont.*

Taxon (Phylum: Class)	Species	Gene Name	Type	Accession No.
Porifera: Calcarea	*Sycon ciliatum*	*SciCHS1*	Type 1	AHX26712.1
	Leucosolenia complicata	*LcoCHS1* *LcoCHS2*	Type 1 Type 1	AHX26702.1 AHX26701.1
Cnidaria: Hexacorallia	*Nematostella vectensis*	*NveCHS1* *NveCHS2*	Type 1 Type 1	EDO41482.1 EDO44996.1
	Stylophora pistillata	*SpiCHS1* *SpiCHS2*	Type 1 Type 1	PFX15170.1 PFX17869.1
	Hydra vulgaris	*HvuCHS1* *HvuCHS2*	Type 1 Type 1	XP_004207525.2 XP_012554922.1
Fungi: Eurotiomycetes	*Aspergillus fumigatus*	*AfuCHS1*	Fungi group	P54267.2
Fungi: Sordariomycetes	*Neurospora crassa*	*NcrCHS1*	Fungi group	P30588.2

* Lophotrochozoan phyla.

3. Results and Discussion

3.1. Data Filtering and Genome Assembly

Using Illumina paired-end sequencing, we generated 39.82 Gb raw reads from the parasitic polychaete *Branchipolynoe onnuriensis*. A stringent quality filter process (with Phred quality scores of 20 or more, see Materials and Methods) was applied; then, a total of 37.58 Gb (94.36%) filtered reads remained. After the filtering process, *de novo* assemblies of the genome sequences, using the software package SPAdes (ver. 3.14.0), were performed. Our initial genome assembly comprised 14,816 contigs, with a total length of 196,561,892 bp. The largest contig was 210,881 bp long, with an N50 length of 12,818 bp. Although the data obtained were insufficient for downstream analysis, since there are no genomic data available in *Brachipolynoe* spp. or parasitic polychaetes, we performed gene prediction to identify CHS protein-coding regions (see Table 2 for general information). .

Table 2. General information from next-generation sequencing to gene annotation in *Branchipolynoe onnuriensis*.

Sequencing	Number of reads before filtering Mean quality score Percentage of $\geq$ Q30 (%) Number of bases (Gb)	263,730,178 35.47 90.95 37.08
Data filtering	Number of reads after filtering	250,683,082
Assembly	Number of contigs (> 10,000 bp) Length of *N*50 (bp) Total length of contigs (bp) Length of the largest contig (bp) GC content (%)	14,816 12,818 196,561,892 210,881 43.71
Gene prediction	Number of predicted genes	353,344

3.2. Gene Prediction and Chitin Synthase Search

Gene structure prediction was conducted using the ab initio method, which yielded 353,344 protein-coding genes. To extract *CHSs* from *B. onnuriensis*, we performed sequence similarity searches and extracted the top five best genes in each group (type 1; groups A, B, C, and D in type 2) and investigated the sequences thoroughly (Table 3). Two genes were identified as type 1; four in groups A, B, and D; and five in group C. The first top-hit gene in each group was assumed to be the *BonCHS* genes belonging to the corresponding group. However, in group B, the third top-hit gene, *g91735.t1*, was considered a candidate *CHS* gene, because the first and second genes, *g38534.t* and *g45117.t1*, belonged to groups D and A, respectively. In addition, phylogenetic analysis for more sensitive identification showed no outlier (Figure 1). Therefore, we determined these five genes as putative *BonCHS* genes.

In addition, the number of genes in each group was determined to be one, except for group C. For example, the first top-hit gene in group A was included in group A, and all three genes belong to groups B, C, and D. However, in group C, the fourth and fifth top-hit genes belong to groups D and A, respectively. Since information about which group the second and third top-hit genes belong to is unknown, we first added two genes to the phylogenetic tree. As a result, both genes were included in the same clade of group C. Thus, the number of genes in group C was determined to be three. Consequently, from the similarity search and phylogenetic analysis, we extracted seven different *CHS* genes from *B. onnuriensis*: *BonCHS1-BonCHS7*.

3.3. Protein Domain Search, Identification of the GT2 Family, and Multiple Sequence Alignments

The seven *BonCHS* genes (*BonCHS1-BonCHS7*) were subjected to predict the domain structures using the SMART web server (http://smart.embl-heidelberg.de, accessed on 23 March 2022). We found that seven BonCHS sequences have chitin_synth_2 domain (Pfam domain: PF03142), except for BonCHS3 and BonCHS5. We supposed that three genes (*BonCHS3*, *BonCHS4*, and *BonCHS5*) in group C were only partially assembled, due to the limitation of Illumina short-read sequencing and lower coverage depth. However, the BLAST searches against the NCBI and UniProt web server showed that *BonCHSs* with the top-hit was *CHS* genes of the lophotrochozoan species, except for *BonCHS6* (Table 4). Furthermore, multiple sequence alignment was performed using 45 amino acid sequences obtained from lophotrochozoans. The two unique motifs, "EDR" (associated with catalytic function) and "QRRRW" (conferring processivity to CHS), were found to be highly conserved in all annelids and mollusks, suggesting their significance in chitin synthesis (Figure 2) [7,32].

Table 3. BLAST result with *E*-value cut-off threshold of $< 1 \times 10^{-50}$ and length > 300 aa.

Database Type		Top Genes	Length (aa)	*E*-Value	Identified Group	Gene Name	Species
Type 1		*g58373.t1*	838	0	Type 1	*BonCHS7*	*Owenia fusiformis*
		g38534.t1	321	2×10^{-51}	Group D		
Type 2	Group A	*g45117.t1*	733	2×10^{-135}	Group A	*BonCHS1*	*O. fusiformis*
		g38534.t1	321	1×10^{-127}	Group D		
		g120019.t1	304	7×10^{-91}	Group C		
		g91735.t1	755	1×10^{-83}	Group B		
	Group B	*g38534.t1*	321	2×10^{-120}	Group D		*Sabellaria alveolata*
		g45117.t1	733	2×10^{-106}	Group A		
		g91735.t1	755	4×10^{-95}	Group B	*BonCHS2*	
		g120019.t1	304	5×10^{-76}	Group C		
	Group C	*g20614.t1*	464	0	Group C	*BonCHS3*	*S. alveolata*
		g120019.t1	304	2×10^{-142}	Group C	*BonCHS4*	
		g86068.t1	472	3×10^{-137}	Group C	*BonCHS5*	
		g38534.t1	321	8×10^{-129}	Group D		
		g45117.t1	733	3×10^{-106}	Group A		
	Group D	*g38534.t1*	321	3×10^{-137}	Group D	*BonCHS6*	*O. fusiformis*
		g45117.t1	733	2×10^{-122}	Group A		
		g91735.t1	755	5×10^{-84}	Group B		
		g120019.t1	304	3×10^{-69}	Group C		

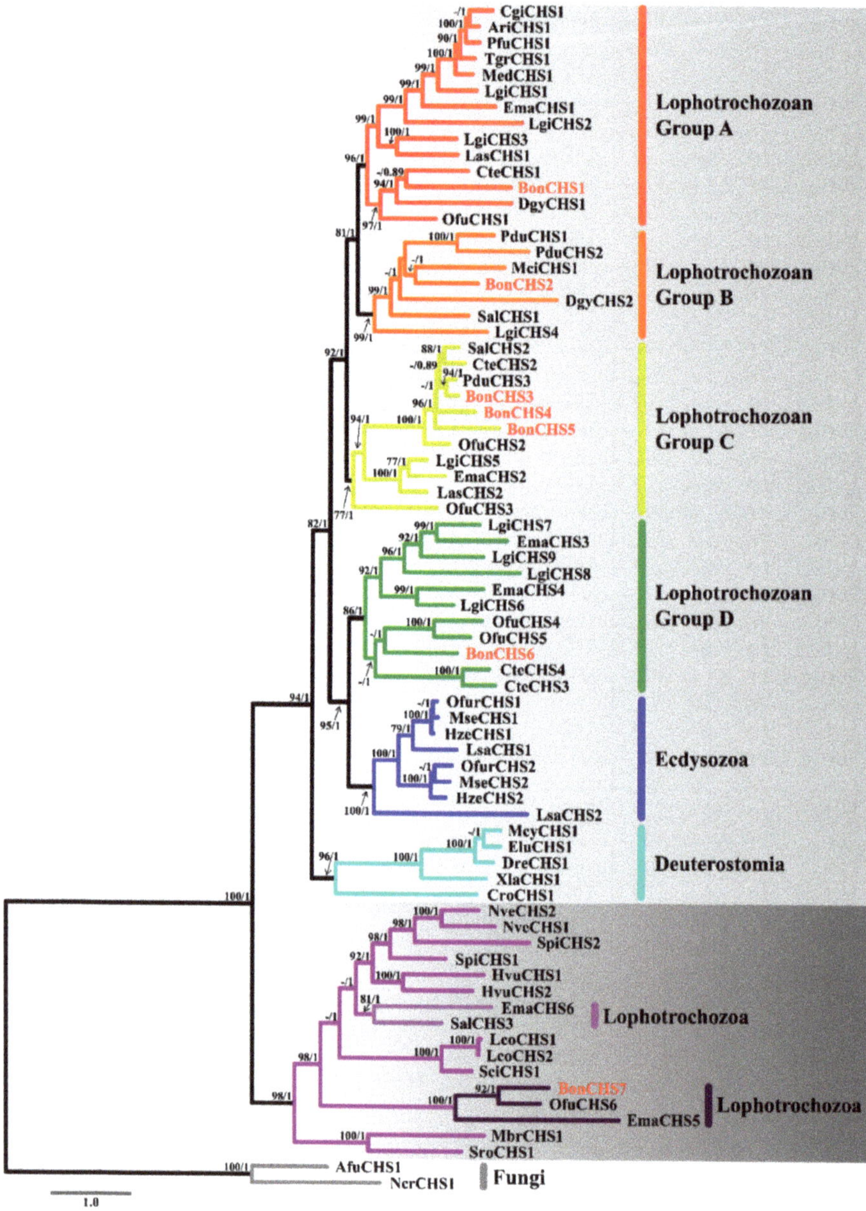

Figure 1. Maximum likelihood (ML) phylogeny of types 1 and 2 chitin synthase (*CHS*) genes: 74 amino acid sequences from 31 metazoans and two fungi. The fungi sequences were used as an outgroup. The gene names are abbreviated and listed in Table 1. *CHS* genes from *Branchipolynoe onnuriensis* are colored in red. Numbers in each node are supporting values and shown only > 70 and > 0.7 from the ML and Bayesian inference. The scale bar is provided at the bottom of the tree and represents the amino acid substitutions per site.

The similarity searches against the CAZy database showed their inclusion in the GT2 family (Table 5). For all genes, the *E*-value was $< 1 \times 10^{-130}$ and their identities ranged from 37.02% to 88.46%. Although BonCHS3 and BonCHS5 were not confirmed by the domain searches, their *E*-values showed 0 and 4.67×10^{-132}, with an identity of 76.46% and 53.22%. Note that our analysis failed to find the "EDR" and "QRRRW" motifs from BonCHS7. Thus, *BonCHS3*, *BonCHS5*, and *BonCHS7* were excluded from the evolutionary patterns.

3.4. Phylogenetic Analysis of Chitin Synthase

In order to confirm the ortholog relationship and understand the molecular evolutionary history, we conducted phylogenetic analysis, including all type 1 and 2 *CHS* genes from the NCBI (Figure 1). As suggested by Zakrzewski et al. (2014) [7], type 1 *CHS* genes generally exist in all metazoans, and *BonCHS7* was found to be closely related to *OfuCHS6* from *O. fusiformis* and *EmaCHS5* from *Elysia marginata*. However, since *E. marginata EmaCHS6* and *S. alveolata SalCHS3* form another clade, type 1 lophotrochozoan CHS is considered a paraphyletic group.

To understand the evolutionary relationship of lophotrochozoan *CHS* genes, we reconstructed a ML phylogenetic tree, with CHS protein sequences from seven annelid species, two gastropods, one polyplacophora, and five bivalves (Figure 3). Five deuterostome sequences were used as an outgroup. Type 2 *CHS* genes mainly consist of four groups (groups A, B, C, and D). In each group, annelid and mollusk clades are clearly separated, with well-supported values of $\geq$ 87% and 1 from the ML and Bayesian inference, which suggests that the lophotrochozoan *CHS* gene duplication event occurred before the divergence of annelids and mollusks [7]. Except for group C, annelid and mollusk *CHS* genes formed a monophyletic clade in all groups. This means that the *O. fusiformis* group C *CHS* gene has undergone a more complex evolutionary process. In the same context, in group C, we found three *BonCHS* genes. Since these genes (*BonCHS3-BonCHS5*) originated from different contigs, they are more likely to result from the gene duplication events, rather than be isoforms. However, not all polychaetes have increased gene copies in group C. For example, two genes were identified from *Platynereis dumerilii* in group B, two and three genes in *O. fusiformis* and *B. onnuriensis* in group C, and two genes in *O. fusiformis* and *C. teleta* in group D, but with no copies in group A. Even for the same taxon, polychaeta, gene duplication did not occur in the same group, which appears to be a species-specific event. Moreover, several *CHS* copies were also found in mollusks (*L. gigantidas* and *E. marginata*). Considering that two types of *CHS* genes with different functions exist in ecdysozoans (component of the exoskeleton and peritrophic matrix), all four different types of CHS of lophotrochozoans may play a different function. Additionally, since *B. onnuriensis* was collected by chance from its host, *Gigantidas vrijenhoeki*, there was a limit to describing their ecological characteristics, except for their habitat and parasitism. However, we obtained evidence of a gene duplication event in group C, and it is best to say that it is due to the two factors mentioned above. To determine the underlying mechanisms and functions of lophotrochozoan enzymes, gene and protein characterization studies are required.

Table 4. The top-hit BLAST results against NCBI and UniProt database.

Database	Query ID	Species	Database ID	Identity (%)	*E*-Value
NCBI	BonCHS1	*Lamellibrachia satsuma*	KAI0208509.1	46.94	2×10^{-156}
	BonCHS2	*Lamellibrachia satsuma*	KAI0242735.1	56.45	1×10^{-92}
	BonCHS3	*Platynereis dumerilii*	AHX26716.1	73.04	0
	BonCHS4	*Platynereis dumerilii*	AHX26716.1	88.09	8×10^{-132}
	BonCHS5	*Sabellaria alveolata*	AHX26711.1	52.52	6×10^{-112}
	BonCHS6	*Homalodisca vitripennis*	KAG8240581.1	64.47	1×10^{-154}
	BonCHS7	*Owenia fusiformis*	CAH1788656.1	51.09	1×10^{-170}

Table 4. *Cont.*

Database	Query ID	Species	Database ID	Identity (%)	*E*-Value
UniProt	BonCHS1	*Capitella teleta*	R7UXD6	46.10	4.3×10^{-161}
	BonCHS2	*Lottia gigantea*	V4B948	44.1	1.7×10^{-121}
	BonCHS3	*Capitella teleta*	R7TXS7	70.30	3×10^{-155}
	BonCHS4	*Capitella teleta*	R7TXS7	87.00	9.3×10^{-135}
	BonCHS5	*Lingula unguis*	A0A1S3IM62	48.10	7×10^{-109}
	BonCHS6	*Bombyx mori*	H9J0C4	66.20	5.7×10^{-153}
	BonCHS7	*Lingula unguis*	A0A1S3IM62	48.10	0

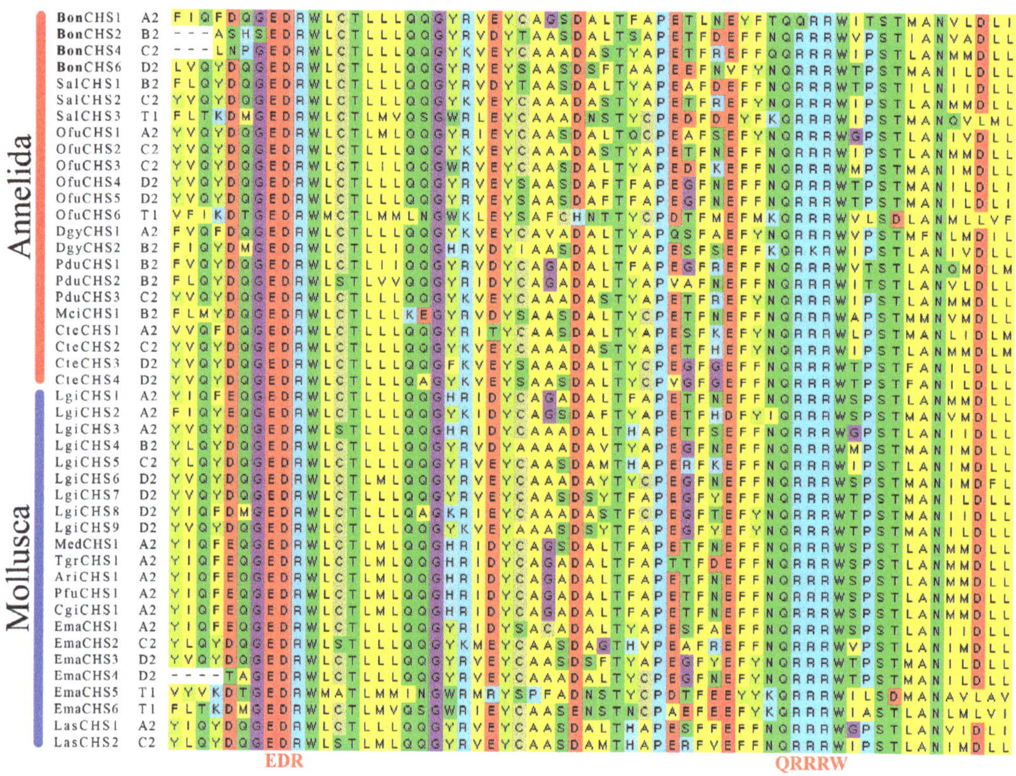

Figure 2. Multiple sequence alignment of CHSs from lophotrochozoan species—23 *CHS* genes from annelids and 22 from mollusks were used. Gene types are marked next to the name (A2, B2, C2, and D2 for groups A, B, C, and D in type 2 and T1 for type 1). Two highly conservative motifs (EDR and QRRRW) are indicated in bold red. The color code is followed by physicochemical properties.

Table 5. The top-hit BLAST results against CAZy database.

Query ID	Species	Database ID	Enzyme Class	Identity (%)	*E*-Value
BonCHS1	*Macandrevia cranium*	AHX26715.1	GT2	42.40	1.54×10^{-154}
BonCHS2	*Myzostoma cirriferum*	AHX26714.1	GT2	37.02	1.71×10^{-141}
BonCHS3	*Platynereis dumerilii*	AHX26716.1	GT2	76.46	0
BonCHS4	*Platynereis dumerilii*	AHX26716.1	GT2	88.46	2.10×10^{-140}
BonCHS5	*Sabellaria alveolate*	AHX26711.1	GT2	53.22	4.67×10^{-132}
BonCHS6	*Bombyx mori*	AFC69002.1	GT2	66.25	4.35×10^{-153}
BonCHS7	*Owenia fusiformis*	AHX26703.1	GT2	50.23	0

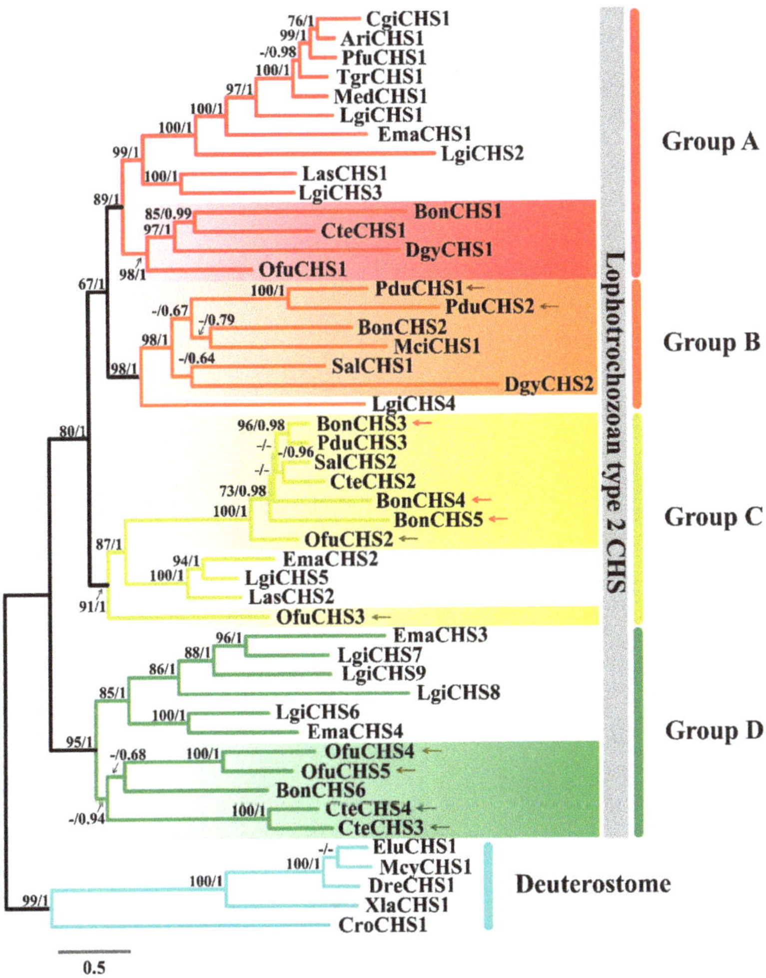

Figure 3. Maximum likelihood phylogeny of chitin synthase genes—48 genes from 15 lophotrochozoan and five deuterostome species are included. The gene names are abbreviated and listed in Table 1. Genes belonging to the polychaetes are colored in each group (red in group A, orange in group B, yellow in group C, and green in group D). Deuterostome sequences are used as an outgroup. In each node, supporting values for ML and Bayesian inference are shown in this order. The nodes supporting values of <60 are indicated with "-". The arrows indicate annelids with several copies (orange for *Platynereis dumerilii*, yellow for *Owenia fusiformis*, green for *Capitellateleta*, and red for *Branchipolynoe onnuriensis*). The clades with the gradient boxes represent the polychaete species in each group. The scale bar represents the amino acid substitutions per site.

4. Conclusions

Chitin, a natural polysaccharide, is the second-largest substance on earth and valuable for many industries. However, compared to the ecdysozoan *CHSs*, which are relatively well-researched, little is known about the lophotrochozoan *CHS* gene. Therefore, in this study, we collected the parasitic polychaete *B. onnuriensis* living in the deep-sea and conducted WGS to investigate the evolutionary aspect of *CHSs*. As a first step toward understanding the role of lophotrochozoan enzymes, we successfully identified seven *CHS* genes (*BonCHS1*-*BonCHS7*) and classified them into five groups. Because of the lower coverage depth and limitation of short-read sequencing, the *B. onnuriensis* genome may have more than seven *CHS* genes. In addition, it is a common phenomenon that lophotrochozoans have several *CHS* genes, especially species living in deep-sea polychaetes. *B. onnuriensis* was found from the host recently; thus, their evolutionary and ecological aspects had remained largely unexplored, except that they survive in the deep-sea and are parasitic. We speculated that the *CHS* gene duplication event might be involved in a harsh environment or parasitic life. Although further research is needed on which tissues the *CHS* gene in group C is expressed and its function, this study suggests the possibility that *CHS* genes in group C are duplicated, which may play a key role in adaptation to parasitic life in harsh environments.

Author Contributions: Conceptualization: S.E.; sample and data collection: S.L.K. and O.H.Y.; data analysis: H.C., S.L.K., M.-K.J., O.H.Y. and S.E.; writing—original draft: H.C., S.L.K., M.-K.J., O.H.Y. and S.E.; writing—review and editing: H.C. and S.E. All authors have read and agreed to the published version of the manuscript.

Funding: This work was supported by the Chung-Ang University Research Scholarship Grants in 2021 and Korea Polar Research Institute (KOPRI, PE22900).

Institutional Review Board Statement: Not applicable.

Informed Consent Statement: Not applicable.

Data Availability Statement: All sequencing data are archived at NCBI Sequence Read Archive (SRA) database, under BioProject number PRJNA817330.

Conflicts of Interest: The authors declare no conflict of interest.

References

1. Zhang, X.; Yuan, J.; Li, F.; Xiang, J. Chitin synthesis and degradation in crustaceans: A genomic view and application. *Mar. Drugs* **2021**, *19*, 153. [CrossRef] [PubMed]
2. Uddowla, M.H.; Kim, A.R.; Park, W.; Kim, H.W. CDNAs encoding chitin synthase from shrimp (*Pandalopsis japonica*): Molecular characterization and expression analysis. *J. Aquac. Res. Dev.* **2015**, *6*, 298. [CrossRef]
3. Abo Elsoud, M.M.; El Kady, E.M. Current trends in fungal biosynthesis of chitin and chitosan. *Bull. Natl. Res. Cent.* **2019**, *43*, 59. [CrossRef]
4. Tharanathan, R.N.; Kittur, F.S. Chitin-the undisputed biomolecule of great potential. *Crit. Reveiws Food Sci. Nutr.* **2003**, *43*, 61–87. [CrossRef]
5. Lee, S.; Kang, H.A.; Eyun, S. Evolutionary analysis and protein family classification of chitin deacetylases in *Cryptococcus neoformans*. *J. Microbiol.* **2020**, *58*, 805–811. [CrossRef]
6. Xin, T.; Li, Z.; Chen, J.; Wang, J.; Zou, Z.; Xia, B. Molecular characterization of chitin synthase gene in *Tetranychus cinnabarinus* (Boisduval) and its response to sublethal concentrations of an insecticide. *Insects* **2021**, *12*, 501. [CrossRef]
7. Zakrzewski, A.; Weigert, A.; Helm, C.; Adamski, M.; Adamska, M.; Bleidorn, C.; Raible, F.; Hausen, H. Early divergence, broad distribution, and high diversity of animal chitin synthases. *Genome Biol. Evol.* **2014**, *6*, 316–325. [CrossRef]
8. Veronico, P.; Gray, L.J.; Jones, J.T.; Bazzicalupo, P.; Arbucci, S.; Cortese, M.R.; di Vito, M.; de Giorgi, C. Nematode chitin synthases: Gene structure, expression and function in *Caenorhabditis elegans* and the plant parasitic nematode *Meloidogyne artiellia*. *Mol. Genet. Genom.* **2001**, *266*, 28–34. [CrossRef]
9. Weiss, I.M.; Schönitzer, V.; Eichner, N.; Sumper, M. The chitin synthase involved in marine bivalve mollusk shell formation contains a myosin domain. *FEBS Lett.* **2006**, *580*, 1846–1852. [CrossRef]
10. Vortsepneva, E.; Tzetlin, A. General morphology and ultrastructure of the radula of *Testudinalia testudinalis* (O.F. Müller, 1776) (Patellogastropoda, Gastropoda). *J. Morphol.* **2019**, *280*, 1714–1733. [CrossRef]
11. Agbaje, O.B.A.; Shir, I.B.; Zax, D.B.; Schmidt, A.; Jacob, D.E. Biomacromolecules within bivalve shells: Is chitin abundant? *Acta Biomater.* **2018**, *80*, 176–187. [CrossRef] [PubMed]
12. Tan, Y.; Hoon, S.; Guerette, P.A.; Wei, W.; Ghadban, A.; Hao, C.; Miserez, A.; Waite, J.H. Infiltration of chitin by protein coacervates defines the squid beak mechanical gradient. *Nat. Chem. Biol.* **2015**, *11*, 488–495. [CrossRef] [PubMed]
13. Picken, L.; Lotmar, W. Oriented protein in chitinous structures. *Nature* **1950**, *165*, 599–600. [CrossRef]
14. Merzendorfer, H. The cellular basis of chitin synthesis in fungi and insects: Common principles and differences. *Eur. J. Cell Biol.* **2011**, *90*, 759–769. [CrossRef] [PubMed]
15. Coutinho, P.M.; Deleury, E.; Davies, G.J.; Henrissat, B. An evolving hierarchical family classification for glycosyltransferases. *J. Mol. Biol.* **2003**, *328*, 307–317. [CrossRef]

16. Braden, L.; Michaud, D.; Igboeli, O.O.; Dondrup, M.; Hamre, L.; Dalvin, S.; Purcell, S.L.; Kongshaug, H.; Eichner, C.; Nilsen, F.; et al. Identification of critical enzymes in the *Salmon louse* chitin synthesis pathway as revealed by RNA interference-mediated abrogation of infectivity. *Int. J. Parasitol.* **2020**, *50*, 873–889. [CrossRef]
17. Luo, Y.-J.; Takeuchi, T.; Koyanagi, R.; Yamada, L.; Kanda, M.; Khalturina, M.; Fujie, M.; Yamasaki, S.-I.; Endo, K.; Satoh, N. The *Lingula* genome provides insights into brachiopod evolution and the origin of phosphate biomineralization. *Nat. Commun.* **2015**, *6*, 8301. [CrossRef]
18. Sun, Y.; Sun, J.; Yang, Y.; Lan, Y.; Ip, J.C.-H.; Wong, W.C.; Kwan, Y.H.; Zhang, Y.; Han, Z.; Qiu, J.-W.; et al. Genomic signatures supporting the symbiosis and formation of chitinous tube in the deep-sea tubeworm *Paraescarpia echinospica*. *Mol. Biol. Evol.* **2021**, *38*, 4116–4134. [CrossRef]
19. Kim, S.L.; Choi, H.; Eyun, S.; Kim, D.; Yu, O.H. A new *Branchipolynoe* (Aphroditiformia: Polynoidae) scale worm from the Onnuri Deep-sea hydrothermal vent field, northern Central Indian Ridge. *Zool. Stud.* **2022**, *61*, 21.
20. Ryu, T.; Kim, J.G.; Lee, J.; Yu, O.H.; Yum, S.; Kim, D.; Woo, S. First transcriptome assembly of a newly discovered vent mussel, Gigantidas vrijenhoeki, at Onnuri Vent Field on the northern Central Indian Ridge. *Mar. Genom.* **2021**, *57*, 100819. [CrossRef]
21. Martin, M. Cutadapt removes adapter sequences from high-throughput sequencing reads. *EMBnet J.* **2011**, *17*, 3. [CrossRef]
22. Bankevich, A.; Nurk, S.; Antipov, D.; Gurevich, A.A.; Dvorkin, M.; Kulikov, A.S.; Lesin, V.M.; Nikolenko, S.I.; Pham, S.; Prjibelski, A.D.; et al. SPAdes: A new genome assembly algorithm and its applications to single-cell sequencing. *J. Comput. Biol.* **2012**, *19*, 455–477. [CrossRef] [PubMed]
23. Gurevich, A.; Saveliev, V.; Vyahhi, N.; Tesler, G. QUAST: Quality assessment tool for genome assemblies. *Bioinformatics* **2013**, *29*, 1072–1075. [CrossRef] [PubMed]
24. Stanke, M.; Morgenstern, B. AUGUSTUS: A web server for gene prediction in eukaryotes that allows user-defined constraints. *Nucleic Acids Res.* **2005**, *33*, W465–W467. [CrossRef]
25. Camacho, C.; Coulouris, G.; Avagyan, V.; Ma, N.; Papadopoulos, J.; Bealer, K.; Madden, T.L. BLAST+: Architecture and applications. *BMC Bioinform.* **2009**, *10*, 421. [CrossRef]
26. Schultz, J.; Copley, R.R.; Doerks, T.; Ponting, C.P.; Bork, P. SMART: A web-based tool for the study of genetically mobile domains. *Nucleic Acids Res.* **2000**, *28*, 231–234. [CrossRef]
27. Cantarel, B.L.; Coutinho, P.M.; Rancurel, C.; Bernard, T.; Lombard, V.; Henrissat, B. The Carbohydrate-Active EnZymes database (CAZy): An expert resource for Glycogenomics. *Nucleic Acids Res.* **2009**, *37*, D233–D238. [CrossRef]
28. Katoh, K.; Misawa, K.; Kuma, K.I.; Miyata, T. MAFFT: A novel method for rapid multiple sequence alignment based on fast Fourier transform. *Nucleic Acids Res.* **2002**, *30*, 3059–3066. [CrossRef]
29. Kalyaanamoorthy, S.; Minh, B.Q.; Wong, T.K.F.; von Haeseler, A.; Jermiin, L.S. ModelFinder: Fast model selection for accurate phylogenetic estimates. *Nat. Methods* **2017**, *14*, 587–589. [CrossRef]
30. Kozlov, A.M.; Darriba, D.; Flouri, T.; Morel, B.; Stamatakis, A. RAxML-NG: A fast, scalable and user-friendly tool for maximum likelihood phylogenetic inference. *Bioinformatics* **2019**, *35*, 4453–4455. [CrossRef]
31. Ronquist, F.; Teslenko, M.; van der Mark, P.; Ayres, D.L.; Darling, A.; Höhna, S.; Larget, B.; Liu, L.; Suchard, M.A.; Huelsenbeck, J.P. MrBayes 3.2: Efficient bayesian phylogenetic inference and model choice across a large model space. *Syst. Biol.* **2012**, *61*, 539–542. [CrossRef] [PubMed]
32. Yu, H.-Z.; Li, N.-Y.; Xie, Y.-X.; Zhang, Q.; Wang, Y.; Lu, Z.-J. Identification and functional analysis of two chitin synthase genes in the common cutworm, *Spodoptera litura*. *Insects* **2020**, *11*, 253. [CrossRef] [PubMed]

Journal of
*Marine Science
and Engineering*

MDPI

Article

Evidence of the Intrusion of the Oceanic Lightfish (*Vinciguerria nimbaria*) into Korean Waters Based on High-Throughput Sequencing of Mixed Fish Eggs

Sung Kim [1,*], Byung-sun Chin [2] and Soon-young Wang [3]

[1] Marine Ecosystem and Biological Research Center, Korea Institute of Ocean Science & Technology, Yeongdo-gu, Busan Metropolitan City 49111, Republic of Korea
[2] Department of Marine Biotechnology, Anyang University, Ganghwa, Incheon 23038, Republic of Korea
[3] Environmental Engineering and Consulting Co., Ltd., Uiwang-si 16006, Republic of Korea
* Correspondence: skim@kiost.ac.kr

Abstract: The oceanic lightfish *Vinciguerria nimbaria* is a mesopelagic species found in tropical and subtropical waters. In this study, we collected a total of 266 mixed fish egg samples from 78 stations in Korean waters of the Northwest Pacific Ocean from January to November 2021, and analysed these samples for *V. nimbaria* using cytochrome c oxidase I (COI) metabarcoding. We detected *V. nimbaria* eggs five times in May and once in August, with 20 *V. nimbaria* eggs to be estimated among the 266 mixed samples, which consisted of 68,844 eggs. To verify the accuracy of the metabarcoding results, two samples consisting of 1 and 6 eggs (diameter, 0.82 ± 0.07 mm; n = 5), respectively, that were identified as *V. nimbaria* were reanalysed using partial COI regions with the Sanger sequencing method. COI sequences obtained using both sequencing methods showed 100% identity in the overlapping regions. The mixed eggs formed one clade with *V. nimbaria* (average pairwise genetic distance, 0.002 ± 0.003; n = 7) in a phylogenetic ML tree based on the mitogenome (2 rRNAs and 13 protein-coding genes) of order Stomiiformes, including partial COIs from the mixed egg samples. The pairwise genetic distances in this clade were smaller than that of Stomiiformes (0.468 ± 0.081), except for *V. nimbaria*. These eggs represent direct evidence of the intrusion or distribution of adult *V. nimbaria*, an unrecorded species, in Korean waters.

Keywords: *Vinciguerria nimbaria*; mixed fish eggs; high-throughput sequencing

1. Introduction

Vinciguerria nimbaria (Jordan and Williams 1895; Stomiiformes: Phosichthyidae), a mesopelagic species found abundantly at depths of 200–400 m in tropical and subtropical oceans [1,2], is distributed in the Western Central, Eastern, and Northwestern Atlantic Oceans [3–5] and Indo-Pacific waters, including the Southeastern Pacific Ocean and South China Sea [6,7] except for Korean waters [8]. *Vinciguerria nimbaria* feeds mainly on copepods and is a major prey of tuna [2,9]; it forms loose schools that occur in clusters during the daytime and dense schools in large aggregations during the night [10], organized through large vertical diel migrations [11].

The total life span of *V. nimbaria* is 6–7 months. Mature females (standard length, >30.6 mm; age, >85 days) spawn pelagic eggs (diameter, 0.65 mm) continuously throughout the year [5]. The lifetime fecundity of *V. nimbaria* is approximately 9000 eggs (109,000 eggs without considering mortality), and the maximum stock egg production of a theoretical cohort occurs at a standard length of 37 mm [5].

Many marine teleost fishes release large quantities of pelagic eggs as a spawning strategy [12,13]. The probability of finding pelagic eggs increases exponentially during the spawning period, exceeding that of adults [14,15]. The geographic distribution of these eggs is critical to locating spawning fish, as well as identifying their spawning grounds and periods. Because fish eggs are extremely sensitive to environmental changes, spawning conditions can be used as ecological indicators of climate change [16].

The identification of fish eggs to the species level using morphological traits alone is difficult because of their high morphological similarity [17,18]. DNA barcode analysis has become increasingly prevalent in egg identification for a limited number of species, including *Lophius litulon* [19], Anguilliformes [20], and *Larimichthys polyactis* [21]. Egg DNA barcoding is also useful for the long-term monitoring of various fish spawning grounds [22,23]. Recently, the intrusion of rare fish that were previously unreported in Korean waters has been detected through DNA metabarcoding of mixed fish eggs based on high-throughput sequencing (HTS) [24].

In this study, we applied DNA metabarcoding to mixed fish egg samples to test its efficacy in monitoring rare species. The *Vinciguerria nimbaria* barcode reads from these samples were confirmed using Sanger sequencing. Our results offer evidence of the distribution of unrecorded adult *V. nimbaria* in Korean waters based on DNA metabarcoding of pelagic fish eggs.

2. Materials and Methods

Fish egg samples were collected from 78 stations in the coastal waters of Korea between January and November 2021. A map of the sampling stations was created using ggOceanMaps software [25] (Figure 1). Egg sampling was conducted using two types of nets (mesh, 300 μm), with mouth diameters of 60 and 80 cm (net towing speed, 1 m/s; net trajectory; vertical, horizontal, and oblique; net towing time, <10 min). We collected a total of 266 specimens containing fish eggs; these were stored in 95% ethanol in the field on the research vessel (10% sample volume in ethanol). A total of 68,844 eggs were extracted from the raw samples under a dissecting microscope (Table 1). The water temperature and salinity of the study area were measured using a Sea-Bird SBE 9 or Sea-Bird SBE 19 Plus V2 instrument (Sea-Bird Electronics, Bellevue, WA, USA).

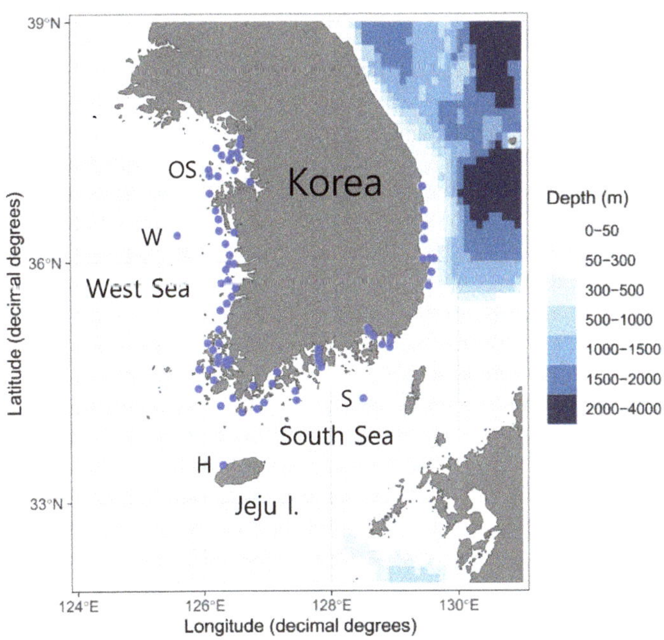

Figure 1. Sampling stations where pelagic fish eggs were collected off the Korean coast.

Table 1. Information on the survey area and high-throughput sequencing data for pelagic fish eggs.

Survey Area	Sampling Date	Sampling Gears	Number of Stations	Sampling Times	Number of Eggs	Total Sequence Reads
H: Hallim of Jeju Is.	26 May 2021	Mouth diameter, 60 or 80 cm; Mesh size, 300 μm	1	5	892	1,198,048
S: EEZ of the southern Korean waters	30 May 2021		1	16	107	4,337,922
W: EEZ of the western Korean waters	18 May 2021		1	15	114	4,025,376
	5 August 2021		1	15	111	3,616,582
OS: Ongjin-Sungap Is. of the western Korean waters	14 May 2021		1	11	42	2,575,872
	3 August 2021		1	27	584	7,532,516
Korean coastal waters	January–November 2021	Mouth diameter, 80 cm; Mesh size, 300 μm	74	177	66,994	56,642,638
Total			78	266	68,844	79,928,954

We performed genomic DNA (gDNA) extraction from one egg to mixed eggs up to 250 ul or less per sample and two adult samples of *Vinciguerria* sp. (Indian Ocean; 4.097° N, 77.350° E) on March 23, 2018 (accession no.: OP983980, OP983981) using a MagListo^{TM} 5M Genomic DNA Extraction Kit (Bioneer, Daejeon, Korea) following the manufacturer's protocols. Prior to gDNA extraction, all samples were photographed using a digital camera. A MiSeq next-generation sequencer (Illumina, San Diego, CA, USA) was used to analyse the cytochrome c oxidase I (COI) barcode region of mixed eggs. Sequencing library construction consisted of two steps: polymerase chain reaction (PCR) and amplicon sequencing [26]. The first PCR was performed using the COI region primers used for species identification, including the MiSeq adapter. The primers used for the first PCR were as follows: forward, TCGTCGGCAGCGTCAGATGTGTATAAGAGACAGGGWACWG-GWTGAACWGTWTAYCCYCC; reverse, GTCTCGTGGGCTCGGAGATGTGTATAAGA-GACAGTAIACYTCIGGRTGICCRAARAAYCA [26,27]. The PCR conditions were: initial denaturation at 95 °C for 3 min; 40 cycles of denaturation at 95 °C for 30 s, annealing at 46 °C for 30 s, and extension at 72 °C for 1 min; and a final extension at 72 °C for 5 min. The products were held at 4 °C.

For Sanger sequencing of two *Vinciguerra* sp. adults (Figure 2) identified followed by Smith and Heems [28], and Nakao [7], one egg (SM310), and six mixed eggs (SM332), PCR was performed using COI fish cocktail primers, as follows: forward, M13(20)-VF2_t1 GTAAACGACGGCCAGTCAACCAACCACAAAGACATTGGCAC and M13(20)-FishF2_t1 GTAAAACGACGGCCAGTCGACTAATCATAAAGATATCGGCAC; and reverse, M13(40)-FishR2_t1 CAGGAAACAGCTATGACACTTCAGGGTGACCGAAGAATCAGAA and M13(40)-FR1d_t1 CAGGAAACAGCTATGACACCTCAGGGTGTCCGAARAAYCARAA [29,30]. The PCR conditions were: initial denaturation at 94 °C for 3 min; 35 cycles of denaturation at 94 °C for 30 s, annealing at 52 °C for 40 s, and extension at 72 °C for 1 min; and a final extension at 72 °C for 4 min. The products were held at 4 °C, and then sequenced using a DNA Analyzer (3730XL; Applied Biosystems, Waltham, MA, USA) with a BigDye Terminator v3.1 Cycle Sequencing Kit (Thermo Fisher Scientific, Waltham, MA, USA).

Secondary PCR was performed for sample classification using the purified product of the first PCR, using the following primers: forward, AATGATACGGCGACCACCGAGATC TACAC-index2-TCGTCGGCAGCGT; and reverse, CAAGCAGAAGACGGCATACGAGAT-index1-GTCTCGTGGGCTCGG [26]. The secondary PCR conditions were: initial denaturation at 95 °C for 3 min; 8 cycles of denaturation at 95 °C for 30 s, annealing at 55 °C for 30 s, and extension at 72 °C for 30 s; and a final extension at 72 °C for 1 min. The products were held at 4 °C. The products of the first and second PCR analyses were purified using a MagListo^{TM} 5M PCR Purification Kit (Bioneer).

Figure 2. *Vinciguerria* sp. (IB18IM 4–8; total length, 32.2 mm) collected in the Indian Ocean on 23 March 2018.

The concentration of the purified second PCR product was measured using a NanoDrop 1000 Spectrophotometer (Thermo Fisher Scientific), and then equal amounts of each sample were mixed. A total of 408 raw FASTQ datasets (≥1–14 segmentation analyses per sample; 68,844 eggs within 266 mixed fish egg samples) were obtained by MiSeq analysis (2 × 301 bp; Illumina), merged using the BBMerge tool (default, [31]), and mapped to the COI region of *V. nimbaria* reference sequences (GenBank accession no. AP012958; complete mitogenome) using Geneious v11.1.5 software.

The merged paired contig reads (MPCRs; >99% identity with *V. nimbaria* COI AP012956) were identified as *V. nimbaria* if the mapping ratio of MPCRs to MMPRs (mean merged paired reads) among sampling stations with >1000 MPRs (merged paired reads) was >80%. To validate the identification accuracy, six complete Stomiiformes genomes, including that of *V. nimbaria* (AP012958), were extracted from GenBank [32]. The six mitogens concatenated 2 rRNAs and 13 protein-coding genes (PCGs), and COI contig reads (including two Sanger sequences) constructed from six mixed egg samples and partial COI sequences from two *Vinciguerria* sp. adults were aligned using ClustalW [33]. The GTR + G + I model (1000 bootstraps) was used to construct an ML (maximum likelihood) tree of aligned sequences with MEGA-X software [34,35].

3. Results

A total of 79,928,854 reads were obtained through HTS of 266 mixed fish egg samples (68,844 eggs; 408 raw FASTQ dataset) collected from 78 stations along the Korean coast. A total of 12,272,357 MPRs were constructed from the 79,928,854 bidirectional raw FASTQ reads. Among the MPRs from the raw HTS reads, 79 samples had 1–9806 MPCRs with >99% identity with *V. nimbaria* COI (AP012958). Only six samples passed the quality control condition for identification as *V. nimbaria*, accounting for at least 1551–25,617 MPRs, was >85.7% ratio of MMPRs to MPCRs (three eggs; SM377) (Table 2). Two MPCRs, from one, single-egg sample (SM310) and one, six-egg mixed sample (SM332), that were identified as the same species were found to be 100% concordant in their overlapping regions according to Sanger sequencing (OP975709, OP975710).

Vinciguerria nimbaria eggs were found off the northwest coast of Jeju Island (H: May, one time) and in the Korea Strait (S: May, three times), south of the Korean Peninsula; and in the West Sea (W: once in May and once in August), west of the Korean Peninsula (Table 2). Eggs were collected using vertical, oblique, and surface net towing methods. Surface net towing was performed twice (Table 2). The water temperature of the surface layer was 16.8–17.4 °C, and the salinity was 35.2–35.5 PSU at the three sampling areas (H, S, and W) in May. In August, the water temperature was 27.8–28.1 °C at the surface and 11.4–11.5 °C at the bottom, and the salinity was 31.8–31.9 PSU at the surface and 32.3–32.4 PSU at the bottom at the W sampling area.

The proportion of MMPRs to MPCRs for two *V. nimbaria* eggs (SM310 and SM332) was 99.4 and 556.7% respectively. The proportion of MMPRs to MPCRs for the other four mixed eggs was 85.7–870.9%. If the ratio of MMPRs to MPCRs was >50%, it was estimated as one egg. There were one to nine *V. nimbaria* eggs per sample, for a total of 20 eggs (Table 2). The diameter of the *V. nimbaria* eggs was 0.71–0.90 mm (0.82 ± 0.07 mm; n = 5) in 95% ethanol.

The eggs were spherical, with no oil globule, and were in the (a–e) early and (f) middle stages of development (Figure 3).

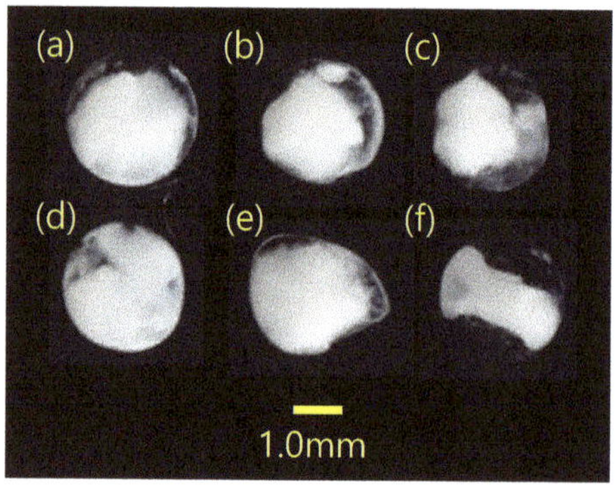

Figure 3. Morphological characteristics of the sampled eggs, with diameters (mm) of (**a**) 0.90, (**b**) 0.84, (**c**) 0.79, (**d**) 0.71, (**e**) no data (not measured because of deformed egg) and (**f**) 0.84. The average egg diameter was 0.82 ± 0.07 mm (n = 5). All eggs were sampled from station W (sample SM332).

All six egg MPCRs formed a phylum with *V. nimbaria* (AP012958), which was used as a standard for species identification in the COI ML tree (Figure 4). Genetic distances within this clade ranged from 0.000 to 0.007 (average, 0.002 ± 0.003; n = 7). This clade was clearly distinguished from the *Vinciguerria* sp. clade, a similar taxon, with a genetic distance of 0.068–0.074 (average, 0.073 ± 0.002). The two lineages of genus *Vinciguerria* were distinctly separated from other taxa in Stomiiformes, with a genetic distance of 0.301–0.660 (average, 0.468 ± 0.081).

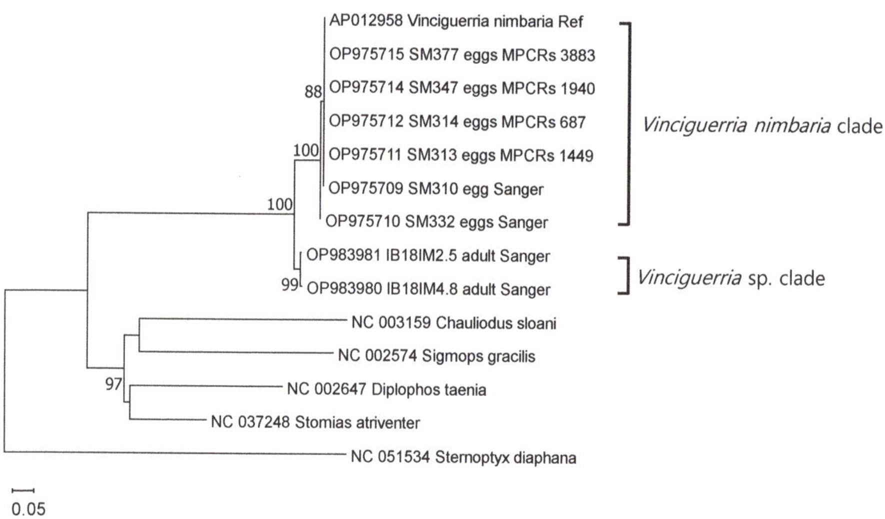

Figure 4. Phylogenetic relationships determined using two rRNAs and 13 protein-coding genes, including partial cytochrome c oxidase I, from six *V. nimbaria* egg merged paired contig reads, two adult *Vinciguerria* sp., and five outgroups as shown in the ML tree. Ref, reference sequence; Sanger, Sanger sequencing; MPRCRs, merged paired read contig reads.

Table 2. Descriptions of reference mapping stations, samples, and merged paired contig reads from mixed pelagic fish egg samples.

Survey Area	Sampling Date	Sample Names	Number of Eggs	Net Towing Trajectory	Net Mouth Diameter (cm)	Bidirectional Raw Reads	Merged Paired Reads	MMPRs (Mean Merged Paired Reads)	MPCRs (Merged Paired Contig Reads)	Ratio of MMPRs to MPCRs (%)	Estimated Number Of *V. Nimbaria* Eggs	GenBank Accession Number	Remark
Hallim of Jeju Is. (33.474° N 126.2863° E)	26 May 2021	SM347	115	vertical	60	270,348	25,617	223	1940	870.9	9	OP975714	
		SM350	6	horizontal	80	274,594	43,341	657	38	5.8	0		
S: EEZ of the southern Korean waters (34.3085° N 128.5057° E)	30 May 2021	SM310	1	Vertical	60	315,646	9869	9869	9806	99.4	1	OP975709	
		SM322	1	oblique	80	269,598	23,508	23,508	1959	8.3	0		
		SM313	11	horizontal	80	300,744	9332	848	1449	170.8	2	OP975711	SS
		SM314	15	horizontal	80	292,366	7290	486	687	141.4	1	OP975712	
		SM312	16	horizontal	80	288,270	12,537	784	28	3.6	0		
OS: Ongjin-Sungap Is. of the western Korean waters (37.0782° N 126.0899° E)	3 August 2021	SM404	24	oblique	60	235,710	10,166	424	33	7.8	0		
		SM390	11	horizontal	80	284,064	1031	94	34	36.3	0		
W: EEZ of the western Korean waters (36.3343° N 125.5795° E)	18 May 2021	SM332	6	oblique	80	274,508	1551	259	1439	556.7	6	OP975710	SS
		SM331	3	horizontal	60	311,818	1084	361	85	23.5	0		
	5 August 2021	SM377	3	oblique	80	259,636	13,600	4533	3883	85.7	1	OP975715	

SS, Sanger sequencing

4. Discussion

Information on the geographical distribution of fish can be obtained from a wide variety of sample types, including eggs, adults, environmental DNA (eDNA; traces of organisms in their habitat), and photographs [23,36–38]. Among these, the easiest samples to attain are eDNA, which can be obtained through water sampling, and fish eggs, which can be collected using small nets or instruments such as the Continuous Underway Fish Egg Sampler [23,39,40]. Unlike eDNA, fish eggs are highly correlated with the distribution of adults during the spawning season, and are therefore useful for identifying spawning grounds through the detection of egg spatiotemporal distribution [41,42]. In this study, we examined the coastal distribution of *V. nimbaria* around the Korean Peninsula by analysing the gDNA of single or mixed fish egg samples by HTS and Sanger sequencing. To date, adults of this species have not been reported in the waters around the Korean Peninsula [8]. The *V. nimbaria* eggs identified by DNA barcoding in this study are the first to be discovered throughout the coastal waters of the Korean Peninsula, as well as in the Pacific Northwest. In a similar case, *Trachipterus jacksonensis*, which had been reported only in the Southern Hemisphere of the Pacific, was found in the East Sea, east of the Korean Peninsula through DNA barcoding of individual eggs [43].

Many marine teleost fishes release large amounts of pelagic eggs for reproduction [12,13]. These eggs have the highest natural mortality rates among teleost life stages. Nevertheless, the likelihood of finding eggs in the early stages of development immediately after spawning is higher than that of adults [14,15]. Thus, most of the eggs collected in this study were in the early stages of development (Figure 3).

Egg size is useful information for selecting DNA barcode target samples when specifying the study area [20,43]. Two species in Korean waters, *Engraulis japonicus* and *Maurolicus japonicus*, were identified as species level without DNA barcoding among lots of fish eggs [15,44,45]. In this study, the average egg size was 0.82 ± 0.07 mm (no oil globule) in 95% ethanol (Figure 3). When fish eggs are stored in 95% ethanol, shrinkage or deformation occurs due to dehydration. However, the *V. nimbaria* eggs collected in this study were approximately 30% larger than those collected in the Atlantic Ocean in a previous study (diameter, 0.65 mm) [5]. Neither the eggs collected in this study nor those collected in the Atlantic had oil globules. Thus, further study is needed to determine whether this large difference in egg size represents intraspecies variation or a potential interspecies difference. Despite these morphological differences, egg morphological traits remain useful for selecting study species.

The length of the DNA barcode region produced via Sanger sequencing was approximately 600 bp, which is a higher resolution than that of the short sequences used in HTS (~313 bp). Large numbers of individual fish egg DNA barcodes require substantial analysis time and cost. An alternative to this costly process is DNA metabarcoding, which is performed on large numbers of nucleotide sequences obtained by HTS based on gDNA extracted from large numbers of mixed fish egg samples. This method is used for various taxa such as zooplankton, and has been applied in eDNA-based fish species exploration [36,37,46] and species analyses of mock mixed fish egg samples or mixed fish larvae [47,48].

In this study, *V. nimbaria* was detected six times through reference mapping of a large number of reads obtained by HTS from a huge quantity of mixed fish egg samples. The two HTS analysis samples, one consisting of a single egg (SM377) and the other of six mixed eggs (SM332), were re-analysed using Sanger sequencing, and the results showed 100% identity of the nucleotide sequences analysed by the two methods (Table 2). These results suggest that our analysis strategy was useful for identifying species from fish eggs.

In fish egg DNA barcode analysis, the accuracy of the reference sequence as the criterion for species identification is important. In this study, the ML tree was created using the mitogenomes (two rRNAs and 13 PCGs) of six fish species from order Stomiiformes, including *V. nimbaria*, the partial COI of two adults of a *Vinciguerria* species, and partial COIs from individual and mixed egg reference mapping contig sequences (including those for

Sanger sequencing) (Figures 2 and 3). In the ML tree, the *V. nimbaria* clade formed a lineage with fish eggs that was closest to the adult *Vinciguerria* species clade. The genetic distance between the two clades was 0.068–0.074 (average, 0.073 ± 0.002). These results confirm that the reference sequence of *V. nimbaria* belongs to the genus *Vinciguerria*, indicating that it may be used as an identifying criterion.

Fertilised eggs of vertebrates such as frogs and mice, including the zebrafish *Danio rerio*, do not replicate mitochondrial DNA during a certain period of development [49–51]. In sardine eggs, the 16S rRNA/18sRNA ratio decreases with embryogenesis, from the early stage to the immediately pre-hatching stage. The ratio of populations occupied by *E. japonicus* and *M. japonicus* in mixed egg samples and the ratio of reads obtained by HTS also showed a significant linear relationship [52]. The ratio of MPCRs to MMPRs for six samples detected *V. nimbaria* were 99.4% and 556.7%, respectively. Based on our results, a total of 20 possible eggs were identified from the ratio of MPCRs in *V. nimbaria* eggs to total MPRs in 151 mixed eggs, including individual eggs. Considering the number of *V. nimbaria* eggs found in three different sea areas, and that this species was found twice within one region after a time interval, and three times in another survey area, there appears to have been a continuous intrusion of *V. nimbaria* adults into the waters around the Korean Peninsula. Therefore, the waters around the Korean Peninsula may be a *V. nimbaria* habitat and the northern limit for spawning grounds based on the direct evidence of *Vinciguerria nimbaria* eggs as shown in the worldwide distribution of *V. nimbaria* (Figure 5).

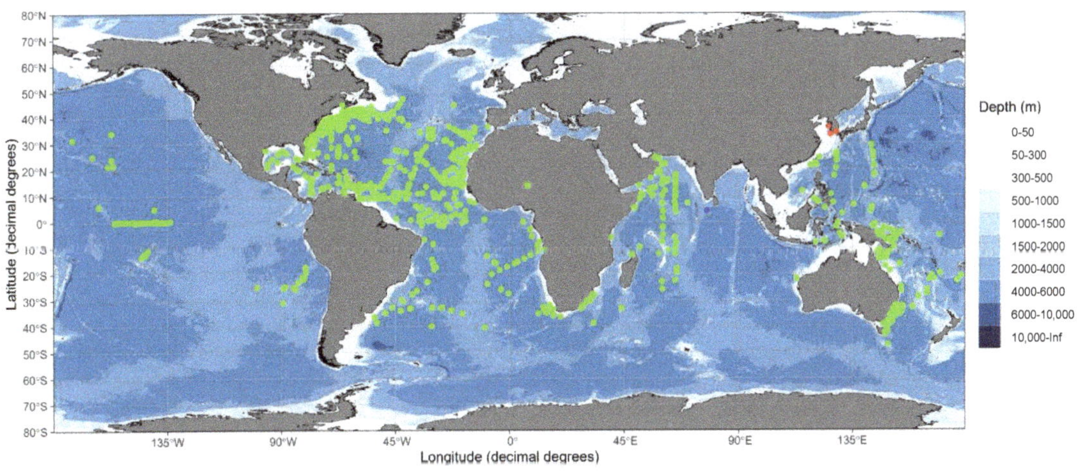

Figure 5. Distribution map (by ggOceanMaps; 24) of *Vinciguerria nimbaria* in tropical and subtropical oceans, showing *V. nimbaria* eggs (red dots) and adult *Vinciguerria* sp. (blue dot) examined in this study, as well as *V. nimbaria* (green dots) from FishBase with permission from [53].

Author Contributions: Conceptualization, S.K.; methodology, S.K.; validation, S.K., B.-s.C. and S.-y.W.; formal analysis, S.K., B.-s.C. and S.-y.W.; investigation, B.-s.C. and S.-y.W.; data curation, B.-s.C. and S.-y.W.; writing— original draft preparation, S.K.; writing—review and editing, S.K., B.-s.C. and S.-y.W.; visualization, S.K.; supervision, S.K., B.-s.C. and S.-y.W.; project administration, S.K., B.-s.C. and S.-y.W.; funding acquisition, S.K., B.-s.C. and S.-y.W. All authors have read and agreed to the published version of the manuscript.

Funding: This research was funded by the Korea Institute of Ocean Science and Technology (PEA0012), the National Research Foundation of Korea (2016R1D1A1B03935737). Additionally, "Development of Advanced Science and Technology for Marine Environmental Impact Assessment" of the Korea Institute of Marine Science & Technology Promotion (KIMST-20210427), and the National Marine Ecosystem Comprehensive Survey supervised by the Korea Marine Environment Corporation, both funded by the Ministry of Oceans and Fisheries. We thank the many researchers who contributed to the field survey and sample analysis of this project and those who operated the survey vessel.

Institutional Review Board Statement: Not applicable.

Informed Consent Statement: Not applicable.

Data Availability Statement: The Sequence data supporting this study's findings are deposited in NCBI/GenBank (https://www.ncbi.nlm.nih.gov/genbank/, accessed on 6 January 2023) under accession numbers OP975709- OP975712, OP975714, OP975715, OP983980, and OP983981.

Conflicts of Interest: The authors declare no conflict of interest.

References

1. Gjøsæter, J.; Kawaguchi, K. A review of the world resources of mesopelagic fish. *FAO Fish. Tech. Pap.* **1980**, *193*, 157.
2. Badcock, J. Photichthyidae. In *Fishes of the North-Eastern Atlantic and the Mediterranean*; Whitehead, P.J.P., Bauchot, M.L., Hureau, J.C., Nielsen, J., Tortonese, E., Eds.; UNESCO: Paris, France, 1984; Volume 1, pp. 318–324.
3. Scott, W.B.; Scott, M.G. Atlantic fishes of Canada. *Can. Bull. Fish. Aquat. Sci.* **1988**, *219*, 731.
4. Claro, R. Características generales de la ictiofauna. In *Ecología de Los Peces Marinos de Cuba*; Claro, R., Ed.; Instituto de Oceanología Academia de Ciencias de Cuba: La Habana, Cuba; Centro de Investigaciones de Quintana Roo: Chetumal, Mexico, 1994; pp. 55–70.
5. Stéquert, B.; Ménard, F.; Marchal, E. Reproductive biology of *Vinciguerria nimbaria* in the equatorial waters of the eastern Atlantic Ocean. *J. Fish Biol.* **2003**, *62*, 1116–1136. [CrossRef]
6. Pequeño, G.; de Peces, C. Lista sistematica revisaday comentada. *Rev. Biol. Mar.* **1989**, *24*, 1–132.
7. Nakabo, T. *Fishes of Japan with Pictorial Keys to the Species*; Tokai University Press: Tokyo, Japan, 2013.
8. Youn, C.H.; Shim, J.H.; Kim, J.J. *Pisces of Korea*; Haksul Information Center: Seoul, Korea, 2021; p. 2148.
9. Lebourges-Dhaussy, A.; Marchal, E.; Menkès, C.; Champalbert, G.; Biessy, B. *Vinciguerria nimbaria* (micronekton), environment and tuna: Their relationships in the Eastern Tropical Atlantic. *Oceanol. Acta* **2000**, *23*, 515–528. [CrossRef]
10. Ménard, F.; Marchal, E. Foraging behaviour of tuna feeding on small schooling *Vinciguerria nimbaria* in the surface layer of the equatorial Atlantic Ocean. *Aquat. Living Resour.* **2003**, *16*, 231–238. [CrossRef]
11. Marchal, E.; Lebourges, A. Acoustic evidence for unusual diel behaviour of a mesopelagic fish (*Vinciguerria nimbaria*) exploited by tuna. *ICES J. Mar. Sci.* **1996**, *53*, 443–447. [CrossRef]
12. Kendall, A.W., Jr.; Ahlstrom, E.H.; Moser, H.G. Early life history stages of fishes and their characters. In *Ontogeny and Systematics of Fishes*; Moser, H.G., Richards, W.J., Cohen, D.M., Fahay, M.P., Kendall, A.W., Richardson, S.L., Eds.; American Society of Ichthyologists and Herpetologists: Kansas, MO, USA, 1984.
13. Leiby, M.M. Life history and ecology of pelagic fish eggs and larvae. *Mar. Plankton Life Cycle Strateg.* **1984**, *6*, 121–140.
14. Houde, E.D. Fish early life dynamics and recruitment variability. *Am. Fish. Soc. Symp.* **1987**, *2*, 17–29.
15. Jung, S.; Hwang, S.D.; Kim, J. Fecundity and growth-dependent mortality of Pacific anchovy (*Engraulis japonicus*) in Korean coastal waters. *Fish. Res.* **2008**, *93*, 39–46. [CrossRef]
16. Dahlke, F.T.; Wohlrab, S.; Butzin, M.; Pörtner, H.O. Thermal bottlenecks in the life cycle define climate vulnerability of fish. *Science* **2020**, *369*, 65–70. [CrossRef] [PubMed]
17. Shao, K.T.; Yang, J.S.; Chen, K.C.; Lee, Y.S. *An Identification Guide of Marine Fish Eggs from Taiwan*; Institute of Zoology, Academia Sinica and Taiwan Power Company: New Taipei, Taiwan, 2001; p. 179.
18. Ikeda, T.; Hirai, A.; Tabata, S.; Onishi, Y.; Mito, S. *An Atlas of Early Stage Fishes in Japan*, 2nd ed.; Okiyama, M., Ed.; Tokai University Press: Tokyo, Japan, 2014; p. 108.
19. Oh, J.; Kim, S. Morphological and molecular characterization of separated pelagic eggs from *Lophius litulon* (Lophiiformes; Lophiidae). *J. Fish Biol.* **2015**, *86*, 1887–1891. [CrossRef] [PubMed]
20. Choi, H.Y.; Oh, J.; Kim, S. Genetic identification of eggs from four species of Ophichthidae and Congridae (Anguilliformes) in the northern East China Sea. *PLoS ONE* **2018**, *13*, e0195382. [CrossRef] [PubMed]
21. Jang, S.H.; Kim, J.K.; Ryu, J.H. First report on the occurrence of eggs of the small yellow croaker *Larimichthys polyactis* from Chilsan-do Island, Jeollanam-do, Korea. *Korean J. Fish. Aquat. Sci.* **2020**, *53*, 650–655.
22. Han, S.H.; Kim, M.J.; Song, C.B. Molecular identification and distribution pattern of fish eggs collected around Jejudo Island. *Korean J. Ichthyol.* **2015**, *27*, 284–292.
23. Harada, A.E.; Lindgren, E.A.; Hermsmeier, M.C.; Rogowski, P.A.; Terrill, E.; Burton, R.S. Monitoring spawning activity in a southern California marine protected area using molecular identification of fish eggs. *PLoS ONE* **2015**, *10*, e0134647. [CrossRef]
24. Choi, H.Y.; Chin, B.S.; Park, G.S.; Kim, S. Evidence of intrusion of a rare species, *Peristedion liorhynchus*, into Korean waters based on high-throughput sequencing of the mixed fish eggs. *Korean J. Ichthyol.* **2022**, *34*, 8–15. [CrossRef]
25. Vihtakari, M. GgOceanMaps: Plot Data on Oceanographic Maps Using 'ggplot2'. R Package Version 1.3.7. Available online: https://mikkovihtakari.github.io/ggOceanMaps/ (accessed on 6 December 2022).

26. Illumina, I. 16S Metagenomic Sequencing Library Preparation. Preparing 16S ribosomal RNA gene amplicons for the illumina MiSeq system. *Microb. Genom.* **2013**, *1*, 28.
27. Leray, M.; Yang, J.Y.; Meyer, C.P.; Mills, S.C.; Agudelo, N.; Ranwez, V.; Boehm, J.T.; Machida, R.J. A new versatile primer set targeting a short fragment of the mitochondrial COI region for metabarcoding metazoan diversity: Application for characterizing coral reef fish gut contents. *Front. Zool.* **2013**, *10*, 34. [CrossRef]
28. Smith, M.M. , *Heemstra, P.C. Smiths' Sea Fishes*; Springer: Berlin/Heidelberg, Germany, 1986.
29. Ward, R.D.; Zemlak, T.S.; Innes, B.H.; Last, P.R.; Hebert, P.D. DNA barcoding Australia's fish species. *Philos. Trans. R. Soc. B Biol. Sci.* **2005**, *360*, 1847–1857. [CrossRef]
30. Ivanova, N.V.; Zemlak, T.S.; Hanner, R.H.; Hebert, P.D. Universal primer cocktails for fish DNA barcoding. *Mol. Ecol. Notes* **2007**, *7*, 544–548. [CrossRef]
31. Bushnell, B.; Rood, J.; Singer, E. BBMerge—Accurate paired shotgun read merging via overlap. *PLoS ONE* **2017**, *12*, e0185056. [CrossRef] [PubMed]
32. Clark, K.; Karsch-Mizrachi, I.; Lipman, D.J.; Ostell, J.; Sayers, E.W. GenBank. *Nucleic Acids Res.* **2016**, *44*, D67–D72. [CrossRef] [PubMed]
33. Thompson, J.D.; Higgins, D.G.; Gibson, T.J. CLUSTAL W: Improving the sensitivity of progressive multiple sequence alignment through sequence weighting, position-specific gap penalties and weight matrix choice. *Nucleic Acids Res.* **1994**, *22*, 4673–4680. [CrossRef] [PubMed]
34. Nei, M.; Kumar, S. *Molecular Evolution and Phylogenetics;* Oxford University Press: New York, NY, USA, 2000.
35. Kumar, S.; Stecher, G.; Li, M.; Knyaz, C.; Tamura, K. MEGA X: Molecular evolutionary genetics analysis across computing platforms. *Mol. Biol. Evol.* **2018**, *35*, 1547–1549. [CrossRef] [PubMed]
36. Bakker, J.; Wangensteen, O.S.; Chapman, D.D.; Boussarie, G.; Buddo, D.; Guttridge, T.L.; Hertler, H.; Mouillot, D.; Vigliola, L.; Mariani, S. Environmental DNA reveals tropical shark diversity in contrasting levels of anthropogenic impact. *Sci. Rep.* **2017**, *7*, 16886. [CrossRef] [PubMed]
37. Boussarie, G.; Bakker, J.; Wangensteen, O.S.; Mariani, S.; Bonnin, L.; Juhel, J.-B.; Kiszka, J.J.; Kulbicki, M.; Manel, S.; Robbins, W.D.; et al. Environmental DNA illuminates the dark diversity of sharks. *Sci. Adv.* **2018**, *4*, eaap9661. [CrossRef]
38. Kim, G.; Song, Y. Identification of freshwater fish species in Korea using environmental DNA technique from the experiment at the freshwater fish ecological learning center in Yangpyeong, Gyeonggi Do. *J. Environ. Impact Assess.* **2021**, *30*, 1–12.
39. Checkley, D.M., Jr.; Ortner, P.B.; Settle, L.R.; Cummings, S.R. A continuous, underway fish egg sampler. *Fish. Oceanogr.* **1997**, *6*, 58–73. [CrossRef]
40. Lelièvre, S.; Verrez-Bagnis, V.; Jérôme, M.; Vaz, S. PCR-RFLP analyses of formalin-fixed fish eggs for the mapping of spawning areas in the Eastern Channel and Southern North Sea. *J. Plankton Res.* **2010**, *32*, 1527–1539. [CrossRef]
41. Fox, C.J.; Taylor, M.; Dickey-Collas, M.; Fossum, P.; Kraus, G.; Rohlf, N.; Munk, P.; van Damme, C.J.G.; Bolle, L.J.; Maxwell, D.L.; et al. Mapping the spawning grounds of North Sea cod (*Gadus morhua*) by direct and indirect means. *Proc. R. Soc. B: Biol. Sci.* **2008**, *275*, 1543–1548. [CrossRef] [PubMed]
42. Tsukamoto, K.; Chow, S.; Otake, T.; Kurogi, H.; Mochioka, N.; Miller, M.J.; Aoyama, J.; Kimura, S.; Watanabe, S.; Yoshinaga, T.; et al. Oceanic spawning ecology of freshwater eels in the western North Pacific. *Nat. Commun.* **2011**, *2*, 179. [CrossRef] [PubMed]
43. Choi, H.Y.; Choi, H.C.; Kim, S.; Oh, H.J.; Youn, S.H. Discovery of pelagic eggs of two species from the rare mesopelagic fish genus *Trachipterus* (Lampriformes: Trachipteridae). *J. Mar. Sci. Eng.* **2022**, *10*, 637. [CrossRef]
44. Kim, J.Y. Relationship Between Anchovy, *Engraulis japonica*, egg and larval density and environmental factors in the eastern waters of Korea. *Korean J. Fish. Aquat. Sci.* **1992**, *25*, 495–500.
45. Kim, S.; Yoo, J.M. Distribution of Eggs and Larvae of *Maurolicus muelleri* in the Thermal Front of the Korea Strait. *Korean J. Ichthyol.* **1999**, *11*, 62–71.
46. Song, C.-U.; Choi, H.; Jeon, M.-S.; Kim, E.-J.; Jeong, H.G.; Kim, S.; Kim, C.-G.; Hwang, H.; Purnaningtyas, D.W.; Lee, S.; et al. Zooplankton diversity monitoring strategy for the urban coastal region using metabarcoding analysis. *Sci. Rep.* **2021**, *11*, 24339. [CrossRef]
47. Kimmerling, N.; Zuqert, O.; Amitai, G.; Gurevich, T.; Armoza-Zvuloni, R.; Kolesnikov, I.; Berenshtein, I.; Melamed, S.; Gilad, S.; Benjamin, S.; et al. Quantitative species-level ecology of reef fish larvae via metabarcoding. *Nat. Ecol. Evol.* **2018**, *2*, 306–316. [CrossRef]
48. Duke, E.M.; Burton, R.S. Efficacy of metabarcoding for identification of fish eggs evaluated with mock communities. *Ecol. Evol.* **2020**, *10*, 3463–3476. [CrossRef]
49. Chase, J.W.; Dawid, I.B. Biogenesis of mitochondria during *Xenopus laevis* development. *Dev. Biol.* **1972**, *27*, 504–518. [CrossRef]
50. Pikó, L.; Taylor, K.D. Amounts of mitochondrial DNA and 132 abundance of some mitochondrial gene transcripts in early mouse embryos. *Dev. Biol.* **1987**, *123*, 364–374. [CrossRef]
51. Artuso, L.; Romano, A.; Verri, T.; Domenichini, A.; Argenton, F.; Santorelli, F.M.; Petruzzella, V. Mitochondrial DNA metabolism in early development of zebrafish (*Danio rerio*). *Biochim. Biophys. Acta BBA Bioenerg.* **2012**, *1817*, 1002–1011. [CrossRef] [PubMed]

52. Choi, H.Y. Diagnosis and Change Prediction of Spawning Areas along the Coasts of the Korean Peninsula Using Pelagic Fish Eggs. Ph.D. Thesis, University of Science and Technology, Daejeon, Republic of Korea, 2021; p. 415.
53. FishBase. World Wide Web Electronic Publication. Froese, R., Pauly D., Eds. Available online: http://www.fishbase.org (accessed on 6 December 2022).

Disclaimer/Publisher's Note: The statements, opinions and data contained in all publications are solely those of the individual author(s) and contributor(s) and not of MDPI and/or the editor(s). MDPI and/or the editor(s) disclaim responsibility for any injury to people or property resulting from any ideas, methods, instructions or products referred to in the content.

MDPI AG
Grosspeteranlage 5
4052 Basel
Switzerland
Tel.: +41 61 683 77 34

Journal of Marine Science and Engineering Editorial Office
E-mail: jmse@mdpi.com
www.mdpi.com/journal/jmse

Disclaimer/Publisher's Note: The title and front matter of this reprint are at the discretion of the Guest Editor. The publisher is not responsible for their content or any associated concerns. The statements, opinions and data contained in all individual articles are solely those of the individual Editor and contributors and not of MDPI. MDPI disclaims responsibility for any injury to people or property resulting from any ideas, methods, instructions or products referred to in the content.

Microwaves Made Simple: Principles and Applications

The Artech House Microwave Library

Introduction to Microwaves by Fred E. Gardiol
Microwaves Made Simple by W. Stephen Cheung and Frederic Levien
Microwave Technology by Erich Pehl
Receivng Systems Design by Stephen J. Erst
Applications of GaAs MESFETs by R.A. Soares, et al.
GaAs Processing Techniques by R.E. Williams
GaAs FET Principles and Technology by J.V. DiLorenzo and D. Khandelwal
Modern Spectrum Analyzer Theory and Applications by Morris Engelson
Microwave Materials and Fabrication Techniques by Thomas S. Laverghetta
Handbook of Microwave Testing by Thomas S. Laverghetta
Microwave Measurements and Techniques by Thomas S. Laverghetta
Principles of Electromagnetic Compatibilty by Bernhard E. Keiser
The Design of Impedance-Matching Networks for Radio-Frequency and Microwave Amplifiers by P.L.D. Abrie
Microwave Filters, Impedance Matching Networks and Coupling Structures by G.L. Matthaei, Leo Young, and E.M.T. Jones
Microwave Engineer's Handbook, 2 vol., Theodore Saad, ed.
Handbook of Microwave Integrated Circuits by R.K. Hoffmann
Microwave Integrated Circuits, Jeffrey Frey and Kul Bhasin, eds.
Computer-Aided Design of Microwave Circuits by K.C. Gupta, R. Garg, and R. Chadha
Microstrip Lines and Slotlines by K.C. Gupta, R. Garg, and I.J. Bahl
Microstrip Antennas by I.J. Bahl and P. Bhartia
Antenna Design Using Personal Computers by David M. Pozar
Microwave Circuit Design Using Programmable Calculators by J. Lamar Allen and Max Medley, Jr.
Stripline Circuit Design by Harlan Howe, Jr.
Microwave Transmission Line Filters by J.A.G. Malherbe
Electrical Characteristics of Transmission Lines by W. Hilberg
Multiconductor Transmission Line Analysis by Sidney Frankel
Microwave Diode Control Devices by Robert V. Garver
Tables for Active Filter Design by Mario Biey and Amedeo Premoli
Active Filter Design by A.B. Williams
Adaptive Electronics by Wolfgang Gaertner
Laser Applications by W.V. Smith
Electronic Information Processing by W.V. Smith
Avalanche Transit-Time Devices, George Haddad, ed.
Gallium Arsenide Bulk and Transit-Time Devices, Lester Eastman, ed.
Ferrite Control Components, 2 vol., Lawrence Whicker, ed.
Microwave Remote Sensing, 3 vol., by F.T. Ulaby, R.K. Moore, and A.K. Fung